Urban Sustainability

The Urban Sustainability Book Series is a valuable resource for sustainability and urban-related education and research. It offers an inter-disciplinary platform covering all four areas of practice, policy, education, research, and their nexus. The publications in this series are related to critical areas of sustainability, urban studies, planning, and urban geography.

This book series aims to put together cutting-edge research findings linked to the overarching field of urban sustainability. The scope and nature of the topic are broad and interdisciplinary and bring together various associated disciplines from sustainable development, environmental sciences, urbanism, etc. With many advanced research findings in the field, there is a need to put together various discussions and contributions on specific sustainability fields, covering a good range of topics on sustainable development, sustainable urbanism, and urban sustainability. Despite the broad range of issues, we note the importance of practical and policy-oriented directions, extending the literature and directions and pathways towards achieving urban sustainability.

The series will appeal to urbanists, geographers, planners, engineers, architects, governmental authorities, policymakers, researchers of all levels, and to all of those interested in a wide-ranging overview of urban sustainability and its associated fields. The series includes monographs and edited volumes, covering a range of topics under the urban sustainability topic, which can also be used for teaching materials.

Ali Cheshmehzangi · Jian Zuo · Ayyoob Sharifi ·
Rongpeng Zhang · Abbas Ziafati Bafarasat ·
Jie Zhao
Editors

Designing Healthy Cities

Integrating Climate-Resilient Urbanism for Sustainable Living

Springer

Editors
Ali Cheshmehzangi
School of Architecture, Design and Planning
The University of Queensland
Brisbane, QLD, Australia

Jian Zuo
School of Architecture and Civil Engineering
University of Adelaide
Adelaide, SA, Australia

Ayyoob Sharifi
The IDEC Institute
Hiroshima University
Hiroshima, Japan

Rongpeng Zhang
School of Architecture and Planning
Hunan University
Changsha, Hunan, China

Abbas Ziafati Bafarasat
Oxford Brookes University
Oxford, Oxfordshire, UK

Jie Zhao
Stuart Weitzman School of Design
University of Pennsylvania
Philadelphia, PA, USA

ISSN 2731-6483 ISSN 2731-6491 (electronic)
Urban Sustainability
ISBN 978-981-96-5537-3 ISBN 978-981-96-5538-0 (eBook)
https://doi.org/10.1007/978-981-96-5538-0

This Springer imprint is published by the registered company Springer Nature Singapore Pte Ltd.
The registered company address is: 152 Beach Road, #21-01/04 Gateway East, Singapore 189721, Singapore

For urban planners, urban designers, and urban specialists who continue to promote urbanism with purpose.

Preface

> *"Health is created and lived by people within the settings of their everyday life; where they learn, work, play, and love."*
>
> —The Ottawa Charter for Health Promotion, from 1986

Cities are at a crossroads of opportunities and crises in an era characterized by rapid urbanization and the growing challenges of climate change. They are both the hubs of vulnerability and the forces behind global advancement, where social injustice, environmental deterioration, and disparities in public health all come together. The core of "*Designing Healthy Cities: Integrating Climate-Resilient Urbanism for Sustainable Living*" is the urgent need to rethink how cities are designed, constructed, and inhabited in light of this duality. This book, which is the second volume in this two-part series, expands on the first volume by exploring the complex relationships among environmental sustainability, public health, and urban design.

This volume emerges at a critical juncture. With more than four billion people living in cities worldwide and still growing, urban systems are under previously unheard-of strain. At the same time, rising temperatures, extreme weather, and the displacement of millions of people are all signs that climate change and its catastrophic impacts are becoming a daily reality rather than a distant threat. These issues are deeply human and affect all facets of life and livelihoods; they are not just technical or infrastructure-related. However, these difficulties also present an unmatched opportunity. Cities have the capacity to spearhead revolutionary change since they are the main hubs for creativity and cooperation. We can design urban areas that actively improve the well-being of their residents while also adapting to the effects of climate change by rethinking them as ecosystems rather than just commercial hubs. By offering a multidisciplinary framework that combines state-of-the-art research, useful tactics, and gripping case studies from around the world, this book aims to stimulate such a paradigm shift.

The book's main proposition is that environmental resilience and public health are inextricably linked. The COVID-19 pandemic has exposed the weaknesses of urban health systems in recent years, highlighting the critical need for spaces that promote mental and physical health. From green spaces that mitigate heat islands and encourage physical activity to sustainable housing that enhances indoor air quality, the book demonstrates how proactive urban design can address the interwoven challenges of climate change and public health. These insights are not just theoretical; they are based on practical applications and provide concrete solutions that are both innovative and useful. More importantly, they are people-centered and they are purpose-driven for better planning and design of our cities and communities worldwide.

The various settings, difficulties, and mistakes of modern urbanism are reflected in the case studies presented in this book. There are many examples, that despite our attempt may not be covered comprehensively in this volume; examples such as those that consider how inclusive and climate-adaptive approaches can promote sustainable development range from flood-resilient communities in Southeast Asia to energy-efficient public housing in Europe and community-driven green infrastructure in Latin America. In order to ensure that no one is left behind in the quest for healthier cities, they also emphasize the significance of participatory governance and the necessity of elevating marginalized voices in the urban planning process. All must happen in the pursuit of healthier cities.

But beyond presenting solutions, this book aims to provoke critical reflection. Why have we, as a global society, allowed cities to become so deeply inequitable? How can we reconcile the demands of economic growth with the constraints of environmental and social justice? And what role do individuals—as citizens, practitioners, and policymakers—play in shaping urban futures? These are not questions with easy answers, but they are questions that must be confronted with courage and creativity.

Designing Healthy Cities is a call to action as much as a guide. It pushes readers to adopt integrated, forward-looking approaches to urban development rather than thinking in silos. This book gives you the information and resources you need to change the world, whether you are an architect, urban planner, legislator, or just someone who is passionate about creating a better society. It challenges you to think of cities as dynamic ecosystems rather than static objects, where resilience and health are the backbone of sustainable living and people and nature coexist peacefully.

The intricacy and pressing nature of the topics covered in this book are well known to us as editors. However, we also have a great deal of hope that human creativity and cooperation will be able to overcome these obstacles. Cities that are healthy and climate-resilient are not a pipe dream; they are attainable if we have the guts to act decisively, think creatively, and collaborate. Inspiring a new generation of urban thinkers and doers to spearhead this change—to create cities that not only endure but flourish in the face of global challenges—is our goal with this volume. Although we do not expect it to be easy or quick, we must all work together to create sustainable urban futures. Let this book be a guide and a call to action for everyone who is dedicated to building cities that care for their people, respect their ecosystems, and act as resilient examples in a world that is changing all the time.

From our five-country locations of Australia, Japan, China, the UK, and the US.

Brisbane, Australia	Ali Cheshmehzangi
Adelaide, Australia	Jian Zuo
Hiroshima, Japan	Ayyoob Sharifi
Changsha, China	Rongpeng Zhang
Oxford, UK	Abbas Ziafati Bafarasat
Philadelphia, USA	Jie Zhao

Acknowledgments We acknowledge everyone's contributions to this exciting book project. We extend our gratitude to all the authors, co-authors, chapter contributors, and collaborators who have shared their expertise, passion, and creativity with us. This sentiment applies to our earlier work as well, where we deeply value the dedication and vision of all contributors who helped to create this book. Such collaborative efforts foster new dialogues, networking opportunities, and a valuable research exchange platform for scholars worldwide.

About This Book

Designing Healthy Cities investigates how urban planning can promote inclusive, climate-resilient, and healthy communities. With cities worldwide battling the twin problems of increasing urbanization and the intensifying effects of climate change, this book offers a relevant, multidisciplinary framework for rethinking urban areas to support equity, well-being, and environmental sustainability.

The book explores the connections between environmental resilience, urban planning, and public health, grounded in state-of-the-art research and enhanced by real-world case studies. It emphasizes how urgent it is for cities to adopt proactive, forward-thinking design strategies in order to both adapt to and lessen the effects of climate change. The book offers practical advice for designing areas that are useful and improve city dwellers' physical and emotional health by fusing sustainable urbanism with human-centered methodologies.

The book emphasizes the necessity of adopting comprehensive, integrated approaches to urban development from a forward-looking standpoint. It urges people to change their perspective from seeing cities only as centers of economic activity to seeing them as living ecosystems that support and care for their residents. By doing this, the book hopes to motivate a new generation of urban leaders and thinkers who are committed to creating cities that coexist peacefully with the environment. It also gives the readers the essential knowledge and resources to help create a sustainable and resilient urban future, whether they are practitioners influencing urban policy or citizens who are enthusiastic about fostering healthier living conditions.

Key Features and Benefits:

- **Holistic Approach to Urban Design:** The book integrates public health, climate resilience, and urban planning into a unified framework, offering actionable strategies for designing inclusive, sustainable cities. This comprehensive perspective helps professionals and communities address interconnected urban challenges effectively.

- **Real-World Case Studies and Best Practices:** Through diverse examples of successful projects worldwide, readers gain practical insights for integrated community-driven and purpose-led initiatives. These real-world applications make complex concepts accessible and replicable.
- **Future-Ready Tools and Frameworks:** Focused on innovation, the book provides guidelines for policymakers, architects, and urban planners to adapt to the rapidly changing dynamics of urbanization and climate impacts. It emphasizes cutting-edge technologies and participatory governance, ensuring cities are prepared for long-term sustainability and resilience.

Contents

Editors and Contributors

About the Editors

Ali Cheshmehzangi is a Professor and Head of the School of Architecture, Design and Planning (ADP) at The University of Queensland, where he leads the School and researches across the fields of architecture, design, and planning/urbanism. He has been in the World's top 2 per cent field leader since 2022, recognised by Stanford University. He is among the top 30 global scholars in the urban sustainability research area. Ali has a wealth of experience, expertise, and a strong commitment to innovation and excellence in the fields of architecture, design, and planning, the three core disciplines of his School at UQ. With a career spanning over two decades, he has made significant contributions to the academic and professional communities, with a focus on sustainable and environmentally conscious design. So far, Ali has published over 500 journal papers, articles, conference papers, book chapters, and reports. He also has 36 other academic books, three of which have received awards at the national, provincial, and municipal levels. He also has received international awards and recognition for his research on urban resilience studies and sustainability research, as well as a Vice-Chancellor's award for his impactful contribution to higher education.

Jian Zuo is a Professor in Sustainable Construction at the School of Architecture and Civil Engineering at the University of Adelaide. His main research interests are related to the sustainable built environment via both technical (e.g. smart technologies) and managerial (e.g. policy intervention and behavioural change) means. In particular, he has undertaken extensive research in construction and demolition waste recycling and management. As the lead researcher, he has been involved in a series of competitive grants in Australia such as Discovery Projects and Linkage Projects funded by the Australian Research Council as well as many other national key projects. He has long been engaged in the study of ecological towns, green buildings, and sustainable development of the construction industry. Published more than 100 papers in top journals, including more than 80 journal papers indexed by SCI and SSCI, Google Scholar indexed more than 28,000, and the H-index is 93. Twelve papers are listed as hot papers/highly cited papers by Incites Essential Science Indicators (ESI).

Ayyoob Sharifi is a Professor at the IDEC Institute, Hiroshima University. He also is a core member of the Network for Education and Research on Peace and Sustainability (NERPS). His research is mainly focused on urban sustainability, resilience, and climate change mitigation and adaptation. He has published extensively on these topics and is ranked among the world's top 1000 influential authors in all fields according to Stanford/Elsevier's Top 2% Scientist Rankings in 2024. Additionally, he received Clarivate's Highly Cited Researcher Award in 2024. Ayyoob actively contributes to global change research programs such as the Future Earth and has served as a lead author for the Sixth Assessment Report (AR6) of the Intergovernmental Panel on Climate Change (IPCC). The goal of his education and research activities is to inform actions toward building sustainable and peaceful communities.

Rongpeng Zhang is a Full Professor and dedicated PhD Supervisor at the School of Architecture and Planning, Hunan University. His primary research pursuits encompass building performance modeling and simulation for intelligent buildings, alongside advanced built environment control for healthy buildings. He was a primary developer of several pivotal modules within the DOE-endorsed EnergyPlus building simulation engine, including VRF-HR, fault detection, and demand response modules. He was a primary developer of Commercial Building Energy Saver (CBES) platform, which obtained the 2019 R&D 100 Award as one of the 100 most significant new technologies in the world. Before joining Hunan University, he led the building science team and conducted a series of research on human-centric built environment control at Well Living Lab, which obtained the 2018 Edison Award. He also served as an assistant professor at Mayo Clinic, and a postdoctoral fellow in the Building Technology and Urban Systems Division at Lawrence Berkeley National Laboratory. He received his Doctorate in Building Performance and Diagnostics from School of Architecture, Carnegie Mellon University and his B.S. and M.S. degrees in Built Environment and Facilities Engineering at Tongji University.

Abbas Ziafati Bafarasat is a Charted Town Planner and Senior Lecturer in Town Planning, School of the Built Environment, Oxford Brookes University, where he also leads the M.Sc. Program in Spatial Planning, and the Certificate in Spatial Planning Studies. Abbas undertook his PhD in Planning and Landscape at the University of Manchester, and his Post Doc at Technical University of Dortmund. His research and professional experience spans strategy and its interdisciplinary applications in enhancing policy and placemaking. Abbas is keen on generative interdisciplinary research and practice systematizing knowledge in confronting emerging issues in sustainable development, with some of his outputs including: (i) proposing an inter-discipline of strategic urban design, (ii) introducing behavioural indicators for sustainability monitoring, (iii) proposing the concept of start-it-yourself urbanism, and (iv) conceptualizing bottom-up agenda-setting.

Jie Zhao is a lecturer of architecture in the Weitzman School of Design at the University of Pennsylvania and Head of Delos Labs and an Executive Vice President at Delos. Delos is a science-based wellness real estate company headquartered in New York City. As the creator of the WELL Building Standard, Delos develops programs and solutions that transform our environments into vehicles for health, well-being, performance, and resilience. Jie leads the global research effort for advancing science and innovation. His research focuses on Human-building Interactions, including healthy buildings, total building performance, occupant behavior in buildings, sensing and controls in buildings, health, comfort, and productivity of building occupants, as well as building energy efficiency and sustainability. He has published 20+ academic articles and serves as reviewers for 10+ top tier academic journals and conferences in the building science field. He frequently contributes to public media as a subject matter expert in healthy buildings, such as Wall Street Journal, CBS News, Forbes, and Money. Jie received his PhD in Building Performance and Diagnostics from Carnegie Mellon University and Master's and Bachelor's degrees from Tongji University.

Contributors

Zaheer Allam School of Architecture, Design and Planning, The University of Queensland, Brisbane, Australia

Abbas Ziafati Bafarasat School of the Built Environment, Oxford Brookes University, Oxford, UK

Ali Cheshmehzangi School of Architecture, Design and Planning (ADP), The University of Queensland, Brisbane, Australia

Cheikh Cisse African Studies Centre Leiden, Leiden University, Leiden, The Netherlands;
Ecole Supérieure des Professions Immobilières, Paris, France

Rahma M. Doheim Al Yamamah University, Riyadh, Saudi Arabia

Alshimaa Aboelmakarem Farag Zagazig University, Zagzig, Egypt

Baojie He School of Architecture, Design and Planning, The University of Queensland, Brisbane, Australia;
Centre for Climate-Resilient and Low-Carbon Cities, Chongqing University, Chongqing, China

Yue Hu School of Architecture and Planning, Hunan University, Changsha, People's Republic of China

Sheng Jiao School of Architecture and Planning, Hunan University, Changsha, People's Republic of China

Ziyi Lin School of Architecture and Planning, Hunan University, Changsha, People's Republic of China

Kourosh Momeni Architecture and Planning Department, Jundi-Shapur University of Technology, Dezful, Iran

Olasunkanmi Habeeb Okunola Vulnerability Assessment, Risk Management and Adaptive Planning Section, United Nations University-Institute for Environment and Human Security, Bonn, Germany

Maryam Roosta Department of Urban Planning and Design, Shiraz University, Shiraz, Iran

Ayyoob Sharifi School of Architecture, Design and Planning (ADP), The University of Queensland, Brisbane, Australia;
The IDEC Institute and Network for Education and Research on Peace and Sustainability (NERPS), Hiroshima University, Hiroshima, Japan

Zhaohui Su School of Public Health, Southeast University, Nanjing, China

Tandis Sulaimani Urban Planning and Management, University of Tehran, Tehran, Iran

Mahdi Suleimany Urban Planning and Management, University of Tehran, Tehran, Iran

Amir Tayyebi Architecture and Planning Department, Jundi-Shapur University of Technology, Dezful, Iran

Roza Vakilinezhad Department of Architecture, Shiraz University, Shiraz, Iran

Xin Wang College of Architecture and Urban Planning, Tongji University, Shanghai, China

Rongpeng Zhang School of Architecture and Planning, Hunan University, Changsha, People's Republic of China

Jie Zhao The Weitzman School of Design, The University of Pennsylvania, Philadelphia, USA;
Delos Labs, Delos, New York City, USA

Wenke Zong School of Architecture and Planning, Hunan University, Changsha, People's Republic of China

Jian Zuo School of Architecture and Civil Engineering, The University of Adelaide, Adelaide, Australia

Tong Zuo Department of Agro-Food Sciences and Technologies (DISTAL), University of Bologna Alma Mater Studiorum, Bologna, Italy

Healthy and Sustainable Living Through Climate-Resilient Urbanism: Moving Forward in Designing Healthy Cities and Communities

Ali Cheshmehzangi, Jian Zuo, Ayyoob Sharifi, Rongpeng Zhang, Abbas Ziafati Bafarasat, and Jie Zhao

Abstract This introductory chapter establishes the book's central focus on designing cities that foster health, equity, and sustainability. It emphasises the urgency of addressing the twin crises of rapid urbanization and climate change by placing the well-being of people and the environment at the forefront of urban development. This chapter presents a unified strategy for tackling the issues facing modern cities by fusing environmental sustainability, urban planning, and public health. With a particular focus on creating healthy and sustainable urban environment, the chapter provides an overview of how climate-resilient urbanism can turn urban areas into ecosystems that are inclusive, adaptable, and conducive to human health. It points out obstacles like resource disparities and policy inertia while suggesting actionable solutions like participatory governance and community-driven practices. The chapter highlights how rethinking cities as places where resilience and well-being are prioritized has the potential to be transformative and catalytic of wider sustainable development outcomes. This introduction provides a conceptual framework that encourages readers to envision and implement strategies for building cities that

A. Cheshmehzangi (✉) · A. Sharifi
School of Architecture, Design and Planning (ADP), The University of Queensland, Brisbane, Australia
e-mail: A.Chesh@uq.edu.au

J. Zuo
School of Architecture and Civil Engineering, The University of Adelaide, Adelaide, Australia

A. Sharifi
The IDEC Institute and Network for Education and Research on Peace and Sustainability (NERPS), Hiroshima University, Hiroshima, Japan

R. Zhang
School of Architecture and Planning, Hunan University, Changsha, China

A. Ziafati Bafarasat
School of the Built Environment, Oxford Brookes University, Oxford, UK

J. Zhao
The Weitzman School of Design, The University of Pennsylvania, Philadelphia, USA

Delos Labs, Delos, New York City, USA

A. Cheshmehzangi et al. (eds.), *Designing Healthy Cities*, Urban Sustainability,
https://doi.org/10.1007/978-981-96-5538-0_1

prosper in the face of global challenges by laying the groundwork for the book's in-depth analysis and practical case studies.

Keywords Healthy cities · Climate resilience · Public health · Sustainable urbanism · Equitable cities · Adaptive design · Urban ecosystems

1 An Overview of Designing Healthy Cities

Unprecedented urbanization and rapid urban growth are features of the twenty-first century, which have been in progress since the mass redevelopments of the 1950s. Over half of the world's population currently lives in cities, and by 2050, that number is expected to rise to almost 70% [1]. In addition to encouraging innovation and economic growth, urbanization has brought with it a number of public health issues [2, 3]. Fast-growing cities often find it difficult to accommodate the basic needs of their citizens [4, 5]. This condition leads to crowded living arrangements, poor sanitation, and limited access to medical care. Vulnerable groups, such as the elderly, children, and low-income groups, are disproportionately impacted by these issues. Urban environments can also exacerbate lifestyle-related health issues [6]. Sedentary behaviours, limited access to green spaces, and the proliferation of processed foods have contributed to rising rates of obesity, diabetes, and cardiovascular diseases. Simultaneously, poor air quality and urban heat islands—caused by vehicular emissions, industrial activities, and densely packed infrastructure—have increased the prevalence of respiratory illnesses and heat-related health risks [7]. Addressing these interconnected challenges requires an integrated approach that prioritizes health as a fundamental component [8] of urban planning and development.

Urban public health is made more complex by the increasing effects of climate change [9]. Cities are especially susceptible to the stresses brought on by climate change because they are centers of economic activity and population density. Urban areas are disproportionately affected by rising temperatures, worsening heat waves, and extreme weather events, which pose serious health risks, cause physical damage, and result in financial losses. Urban flooding, which is made worse by poor drainage systems and the disappearance of natural buffers like wetlands, endangers public health right away by contaminating drinking water supplies and spreading waterborne illnesses [10, 11]. In a similar vein, protracted droughts can deplete water supplies, jeopardizing food security and hygiene [12, 13]. These issues frequently affect marginalized communities the most, emphasizing the necessity of inclusive and equitable climate adaptation measures.

The necessity of rethinking cities as resilient ecosystems is highlighted by the relationship between urban vulnerabilities and climate change [14]. In addition to promoting environmental sustainability, adaptive strategies like strengthening green infrastructure, enhancing building designs, and encouraging community-led solutions can lessen the negative health effects of climate change.

1.1 The Urgent Need for a Paradigm Shift

A fundamental rethinking of urban systems is required in light of the challenges posed by urbanization and climate change [15]. Prioritizing health and well-being must replace traditional city planning, which is frequently centered on infrastructure efficiency and economic growth [16]. According to this paradigm shift, cities are ecosystems that promote health and have a close relationship between environmental and human health [17]. Creating settings that promote active lifestyles, fair access to resources, and improved social cohesion are all part of designing healthy cities. Urban parks, green roofs, and community gardens are examples of nature-based solutions that can be incorporated to enhance air quality, lessen heat stress, and promote mental health [18–20]. In a similar vein, environmentally friendly transportation systems that emphasize cycling, walking, and public transportation can lower pollution and promote healthier living.

Resilience must be incorporated into city governance systems as well [21]. Developing inclusive and adaptable urban environments requires collaborative urban planning that involves a range of stakeholders, including citizens, researchers, policymakers, and community leaders [22, 23]. We can address the interrelated issues of urbanization and climate change while promoting sustainable living for coming generations by converting cities into ecosystems that promote health.

This chapter sets the stage for exploring the strategies, innovations, and interdisciplinary approaches needed to design cities that not only withstand the challenges of the future but also thrive as models of health and sustainability. Thus, this chapter lays the foundation for examining the tactics, inventions, and multidisciplinary approaches required to create cities that not only survive the challenges of the future but also flourish as sustainable and health-modelling examples.

2 Understanding Climate-Resilient Urbanism

In order to ensure that cities can endure, adjust to, and recover from the complex effects of climate change while promoting long-term sustainability and inclusivity, climate-resilient urbanism is a developing paradigm [24, 25]. Principles like flexibility, adaptability, self-organisation, inclusivity, and foresight are all included in this idea. A city that is climate resilient can mitigate chronic stressors like rising temperatures, water scarcity, and ecosystem degradation in addition to being ready for acute shocks like hurricanes and floods.

A key component of climate resilience is adaptability, which calls on cities to constantly change their community practices, policies, and infrastructure in response to new threats. Urban areas can become more resilient to climate variability and energy demands by incorporating renewable energy systems and retrofitting buildings to use less energy or improve energy efficiency [26–28]. Robustness highlights the necessity of infrastructure that can withstand harsh environments, such as transportation networks that are resistant to flooding or construction materials that can withstand high temperatures [29, 30]. All urban dwellers' needs are met by resilience-building initiatives when inclusivity is maintained, especially those of marginalized communities that are disproportionately impacted by climate crises [31–33]. In order to foresee future difficulties and reduce vulnerabilities, foresight also entails proactive planning and risk assessment.

Climate-resilient urbanism has significant ramifications for urban areas. By lowering damage costs, guaranteeing business continuity, and encouraging innovation, these tactics not only benefit the environment but also advance economic stability [34–36]. Socially, by safeguarding vulnerable groups and facilitating inclusive decision-making, resilient cities improve equity and public safety. By incorporating nature-based solutions, like urban forests and green corridors, which improve biodiversity and offer vital ecosystem services, they also aid in ecological regeneration.

It is becoming more widely acknowledged that addressing climate change in urban settings requires a connection between sustainability and public health [37]. Frameworks for urban planning that put sustainability and health first can greatly reduce the negative effects of climate change while improving people's quality of life in general. Recognizing the interconnectedness of the environment, health, and socioeconomic systems, climate-resilient urbanism promotes comprehensive strategies that address all three aspects at once.

Promoting clean air and water plays a crucial role in this integration. High pollution levels in urban areas worsen cardiovascular and respiratory conditions, which are made worse by heat waves and rising temperatures. Urban heat islands can be decreased and air quality improved by implementing planning strategies like establishing low-emission zones, improving public transit, and growing green spaces [38]. Resilient water management systems, such as rainwater collection and sophisticated wastewater treatment, guarantee sustainable and safe water access while lowering the risk of waterborne illnesses.

Benefits to mental and physical health are also essential to sustainable urban planning. Well-planned pedestrian and bicycle infrastructure in walkable cities promotes active lifestyles, which lowers obesity and associated health issues. Furthermore, having access to urban green spaces enhances mental health by reducing stress and anxiety that are frequently associated with urban living and uncertainties related to climate change. Sustainability goals are in line with public health priorities when health metrics are incorporated into urban design procedures, such as evaluating possible health effects during zoning and infrastructure development.

Additionally, food security and nutrition are prioritized in frameworks for climate-resilient urban planning. Local food resilience is improved by urban farming projects, community gardens, and food distribution networks that lessen dependency on lengthy supply chains. These actions address food inequities and malnutrition among urban populations in addition to reducing carbon footprints and promoting sustainability. Cities can build more resilient, equitable, and long-term well-being-promoting environments by integrating sustainability [39] and health as fundamental components.

2.1 *Rethinking Urban Design Strategies and Solutions: Innovative Approaches to Adaptive and Inclusive*

The advancement of climate resilience depends heavily on innovative urban design, which provides inclusive and flexible answers to the hitherto unseen problems caused by climate change [40–42] Using dynamic, adaptable strategies that can react to changing social, economic, and environmental circumstances is a key component of adaptive city planning. For example, modular infrastructure systems offer a workable solution to future urban demand uncertainty because they can be scaled or altered over time.

Innovative urban design is characterized by nature-based solutions, which incorporate ecological systems into the urban fabric to improve resilience [43]. Permeable pavements, green roofs, and wetland restoration initiatives are a few examples that reduce flooding, enhance air quality, and promote biodiversity. In addition to addressing particular climate risks, these solutions develop multipurpose urban areas with ecological, aesthetic, and recreational uses. Similar to this, smart technologies like real-time monitoring systems and predictive climate modelling help cities foresee hazards and maximize resource utilization, guaranteeing effective and adaptable urban management.

By assigning top priority to diverse populations' needs, inclusive city planning makes sure that resilience initiatives benefit all city dwellers equally. In order to develop solutions that take into account local priorities and expertise, community members must actively participate in planning and decision-making processes through participatory design processes. Given that marginalized communities frequently suffer the most from the effects of climate change, this strategy is especially crucial for addressing climate justice. For instance, inclusive housing policies that offer low-income residents affordable, energy-efficient homes improve their living conditions while lowering their exposure to extreme weather.

Another area where inclusivity and innovation meet is in urban mobility. In order to lower greenhouse gas emissions and improve connectivity, climate-resilient transportation systems give priority to environmentally friendly forms of transportation like walking, bicycling, and public transportation. These systems are made to be accessible, meeting the needs of families, senior citizens, and people with disabilities. Cities are further guaranteed to stay operational and connected during climate disruptions through the integration of multi-modal hubs and seamless transit networks.

The goal of rethinking urban design is to promote social and cultural resilience in addition to physical infrastructure. The ability of the community as a whole to deal with climate challenges is strengthened by creating networks of support, funding education and awareness initiatives, and honouring regional heritage and identity. Cities can move toward a future that is not only resilient but also prosperous and equitable [44] in the face of climate change by adopting innovative, inclusive, and adaptive strategies.

3 Barriers to Progress and Pathways Forward

Even though the necessity of health-centered urban development is becoming more widely acknowledged, a number of obstacles stand in the way of advancement. One significant issue is policy inertia, which occurs when modern frameworks are unable to adapt to creative, health-focused urban design concepts. Governments frequently function within inflexible frameworks that are sluggish to adjust to new issues like rising health disparities and climate change. The lack of integrated policies that link sustainability, urban planning, and health exacerbates fragmentation and makes it challenging to take coordinated action.

The implementation of healthy urban designs is further hampered by a lack of resources and limitation in relevant skills. Many local governments struggle to create or maintain sustainable urban infrastructures due to a lack of resources, technical know-how, and capacity. Due to resource scarcity, immediate financial gains are frequently given precedence over long-term environmental and health advantages. For example, despite green infrastructure projects' demonstrated advantages in lowering urban heat and encouraging physical activity, cost-cutting measures may cause them to be neglected. Lastly, unequal access to resources, services, and decision-making procedures seriously hinders the growth of a healthy urban environment. Urban problems like substandard housing, restricted access to green areas, and susceptibility to environmental hazards frequently fall most heavily on marginalized communities. In addition to fair resource distribution, addressing these disparities calls for inclusive procedures that elevate marginalized voices.

3.1 Participatory Governance: Engaging Communities in Shaping Urban Environments

By putting communities at the center of urban development processes, participatory governance provides a vital means of overcoming these obstacles. Involving citizens ensures that interventions meet the specific needs of various populations by fostering a sense of accountability and ownership. To bridge the gap between the public and policymakers, co-design workshops and public consultations, for instance, can enable locals to offer insights into neighbourhood-level problems and solutions [45–47].

Furthermore, participatory methods improve urban governance's openness, transparency, and credibility. People are more likely to support and maintain initiatives that are put into place when they actively participate in influencing their surroundings. Top-down and bottom-up tactics are frequently combined in successful participatory models, which increase community-driven solutions while utilizing government assistance. Communities can choose how to allocate funds through initiatives like participatory budgeting, which helps them match resources to their most urgent needs. Additionally, technology has the potential to revolutionize participatory governance. Wider and more inclusive engagement is made possible by digital platforms for citizen feedback and decision-making tools. Social media campaigns or interactive urban planning apps, for instance, can reach a variety of demographics, including those who have traditionally been shut out of traditional governance procedures.

3.2 Actionable Strategies: Practical Steps for Implementing Climate-Resilient, Health-Centered Designs

Cities must implement practical solutions that remove current obstacles and advance sustainability and equity in order to advance climate-resilient, health-centered urbanism. Health outcomes should be given equal weight to economic and environmental objectives in integrated urban policies. Zoning changes, for example, can encourage mixed-use projects that improve social cohesion, encourage walkability, and shorten commutes. Another sensible step is to invest in natural solutions. In addition to reducing climate risks, urban forests, green roofs, and wetlands improve both physical and mental health. In the long term, these interventions are cost-effective because they promote healthier communities and lower healthcare costs.

To overcome policy fragmentation and resource constraints, cross-sectoral cooperation is crucial. In order to ensure that interventions are both creative and practical, partnerships between public health agencies, urban planners, stakeholders from the private sector, and community organizations can pool resources and expertise. Collaborations among community organizations, private sector stakeholders, urban planners, and public health agencies can share resources, exchange knowledge, and promote creativity. These collaborations foster trust and encourage community involvement in addition to improving interventions' inventiveness and viability. Such collaboration makes it possible to create inclusive, sustainable solutions that meet short-term demands while fostering long-term resilience and adaptability by combining insights from multiple disciplines.

4 Toward a Vision of Resilient and Healthy Cities

As the intertwined challenges of urbanization, climate change, and public health push the existing urban systems to their limits, building resilient and healthy cities is becoming an increasingly important global priority. Public health, economic viability, social equity, and climate resilience must all be incorporated into urban planning and design in order to create a sustainable urban living vision. By providing readers with actionable insights drawn from successful international case studies, this book seeks to bridge the gap between aspiration and action.

Lessons learnt from global cities highlight that resilience is a contextual and iterative process rather than a one-size-fits-all solution [38]. Examples of how localized strategies can have significant effects range from Medellín's revolutionary public transportation initiatives that improved social cohesion to Copenhagen's creative use of green infrastructure to mitigate urban flooding. Cities aiming for a healthier future can learn from Barcelona's introduction of "superblocks" to improve walkability and lower air pollution and Singapore's incorporation of biophilic design principles into its urban fabric. These international achievements demonstrate that the foundation of significant change is cooperation, creativity, and community involvement.

The book's case studies go beyond inspiration to offer practical lessons. They showcase how various stakeholders, including governments, the commercial sector, and grassroots groups, can cooperate to manage complexity and provide solutions that put resilience and well-being first. The policies, design approaches, and governance frameworks that have contributed to success in various urban contexts will become clearer to readers as they examine these examples. The ultimate goal of this book is to empower cities all over the world to rethink their potential, transforming obstacles into opportunities and creating more inclusive, healthy, and future-ready urban environments. Let these global realizations act as a spur for group efforts to build cities that coexist peacefully with the environment and their citizens.

Imagining a prosperous urban future necessitates audacious tactics based on resilience, sustainability, and inclusivity as cities confront previously unheard-of difficulties. The combination of public health emergencies, rapid urbanization, socioeconomic disparities, and climate change necessitates innovative strategies that transcend traditional urban planning paradigms. Cities must embrace creative solutions that combine environmental, social, and economic priorities if they are to prosper in the face of these global challenges.

Adaptive urbanism or designing cities to change dynamically in response to shifting conditions, is a key component of this vision. Green infrastructure—such as permeable surfaces, wetlands, and urban forests—not only mitigates climate impacts but also improves community well-being. Cities like Rotterdam have shown how using natural solutions can both create lively public spaces and fight rising sea levels [48]. Similarly, Curitiba's innovative bus rapid transit (BRT) system is a prime example of how innovative mobility solutions can improve accessibility and lower carbon footprints [49]. The capacity to promote inclusivity and economic resilience is also essential for a prosperous future [50]. To develop smart urban systems that maximize resource use and enhance service delivery, cities must take advantage of innovation and technology. Cities like Singapore can improve urban liveability and achieve energy efficiency with the aid of smart grids, predictive analytics, and IoT-enabled infrastructure. To guarantee that the advantages are distributed among all groups, technological developments must be combined with equity-focused legislation.

Transformation of governance and collaboration are equally important. Participatory decision-making involving a range of stakeholders, including local residents, community organizations, policymakers, and urban planners, is essential to creating thriving cities. In addition to strengthening the social fabric, this inclusivity promotes a sense of responsibility and ownership. Medellín's focus on participatory urbanism provides a compelling illustration of how communities can work together to develop solutions that are tailored to their own needs and goals. Ultimately, prosperous cities are those that actively shape their futures in addition to responding to global challenges. Cities can forge a route toward resilience, sustainability, and vibrancy by adopting adaptive design, promoting equity, utilizing innovation, and placing a high value on teamwork. These tactics, which are based on international case studies [51], provide a roadmap for cities to thrive in the face of hardship and guarantee a time when urban areas and their people coexist peacefully.

4.1 *Call to Action: A Necessity for Designing Healthy Cities and Communities*

There has never been a more pressing need to reinvent urban living as cities around the globe deal with growing difficulties [52]. This book serves as a rallying cry for citizens, policymakers, planners, architects, and urban leaders to step up and take an active role in creating communities that are healthier and more resilient. Our cities' future rests on the combined efforts of those who hope for not only survival but also prosperity in the face of hardship.

In order to direct interventions and track results, cities need to give data-driven decision-making top priority [53]. Priority areas for action can be determined and progress can be monitored over time by utilizing geospatial analysis and health impact assessments. Existing science-informed action and evaluation healthy community framework has been developed and utilized in practice [54]. In order to address disparities and climate risks at the same time, heat maps that highlight urban heat islands, for example, can direct tree-planting campaigns in neighbourhoods that are at risk. Cities can create healthier, more climate-resilient urban environments by tackling policy inertia, resource limitations, and disparities through participatory governance and workable solutions. To ensure that the advantages of urban transformation are distributed fairly, these initiatives call for consistent dedication and cooperation from all societal levels.

Our capacity for audacious thought, teamwork, and responsible innovation holds the answers. Every action counts, from creating fair public areas to putting in place environmentally friendly transit and embracing green infrastructure. This book emphasizes that no contribution is too small and provides readers with global insights and practical strategies to empower and inspire them. While citizens can support grassroots projects and push for laws that put health, well-being, and inclusivity first, leaders can set lofty objectives and promote systemic change.

Lastly, it is important to note that we can build more adaptable, just, and lively cities by taking inspiration from international achievements and applying it to local settings. Allow this book to be your roadmap and source of inspiration, igniting thoughts, encouraging teamwork, and strengthening our shared resolve to create urban spaces that support the well-being of people and the environment.

References

1. World Bank (n.d.) Inclusive cities. Available from: https://www.worldbank.org/en/topic/inclusive-cities
2. Cheshmehzangi A (2020) Reflection on early lessons for urban resilience and public health enhancement during the COVID-19. Health 12(10):1390
3. Sukanya R, Tantia V (2023) Urbanization and the impact on economic development. In: New perspectives and possibilities in strategic management in the 21st century: between tradition and modernity. IGI Global, New York, pp 369–408
4. Cabaj M (2012) Cities reducing poverty: how vibrant communities are creating comprehensive solutions to the most complex problems of our times. BPS Books, New York
5. Cheshmehzangi A (2021) Urban health, sustainability, and peace in the day the world stopped. Springer, Singapore
6. World Health Organization (2024) Achieving well-being: a global framework for integrating well-being into public health utilizing a health promotion approach. World Health Organization, New York
7. Ziafati Bafarasat A, Cheshmehzangi A, Ankowska A (2023) A set of 99 healthy city indicators for application in urban planning and design. Sustain Dev 31(3):1978–1989
8. Jiang X, Ye D, Lan W, Luo Y (2024) Epidemic, urban planning and health impact assessment: a linking and analyzing framework. Buildings 14(7):2141
9. Sharifi A, Pathak M, Joshi C, He BJ (2021) A systematic review of the health co-benefits of urban climate change adaptation. Sustain Cities Soc 74:103190
10. Avni RD (2020) Status and causes of deterioration of wetlands in the Eastern Riftvalley, case study of lake Nakuru-Kenya. Doctoral dissertation, University of Nairobi
11. Muriithi MW (2014) An assessment of human-induced impacts on urban wetlands: a case study of the Nairobi dam. Doctoral dissertation, University of Nairobi
12. Jain S, Srivastava A, Khadke L, Chatterjee U, Elbeltagi A (2024) Global-scale water security and desertification management amidst climate change. Environ Sci Pollut Res 2024:1–25
13. Nairizi S (2017) Drought and water scarcity. International Commission on Irrigation and Drainage, New Delhi
14. Sharifi A (2022) Cities in the context of global change: challenges and the need for smart and resilient cities. In: Resilient smart cities: theoretical and empirical insights. Springer, Cham, pp 3–16
15. McPhearson T, Haase D, Kabisch N, Gren Å (2016) Advancing understanding of the complex nature of urban systems. Ecol Ind 70:566–573

16. Cheshmehzangi A (2021) Revisiting the built environment: 10 potential development changes and paradigm shifts due to COVID-19. J Urban Manag 10(2):166–175
17. Sharifi A (2023) Resilience of urban social–ecological–technological systems (SETS): a review. Sustain Cities Soc 2023:104910
18. Baró F, Camacho DA, Perez del Pulgar C, Ruiz-Mallén I, García-Serrano P (2022) Nature-based climate solutions in European schools: a pioneering co-designed strategy towards urban resilience. In: Urban resilience to the climate emergency: unravelling the transformative potential of institutional and grassroots initiatives. Springer, Cham, pp 125–146
19. Conti ME, Battaglia M, Calabrese M, Simone C (2021) Fostering sustainable cities through resilience thinking: the role of nature-based solutions (NBSs): lessons learned from two Italian case studies. Sustainability 13(22):12875
20. Snep RP, Voeten JG, Mol G, Van Hattum T (2020) Nature based solutions for urban resilience: a distinction between no-tech, low-tech and high-tech solutions. Front Environ Sci 8:599060
21. Boyd E, Juhola S (2015) Adaptive climate change governance for urban resilience. Urban Stud 52(7):1234–1264
22. Chu E, Anguelovski I, Carmin J (2016) Inclusive approaches to urban climate adaptation planning and implementation in the Global South. Clim Policy 16(3):372–392
23. Wolfram M (2016) Conceptualizing urban transformative capacity: a framework for research and policy. Cities 51:121–130
24. Rahman MA, Hossain MZ, Rahaman KR (2023) Climate urbanism as a new urban development paradigm: evaluating a city's progression towards climate urbanism in the global south. Climate 11(8):159
25. World Health Organization (2010) Hidden cities: unmasking and overcoming health inequities in urban settings. World Health Organization. Centre for Health Development, New York
26. Aristizabal S, Porter P, Clements N, Campanella C, Zhang R, Hovde K, Lam C (2019) Conducting human-centered building science at the well living lab. Technol | Archit + Des 3(2):161–173
27. Ding Z, Niu J, Liu S, Wu H, Zuo J (2020) An approach integrating geographic information system and building information modelling to assess the building health of commercial buildings. J Clean Prod 257:120532
28. Hansen A, Williamson T, Pisaniello D, Bennetts H, Van Hoof J, Arakawa Martins L, Visvanathan R, Zuo J, Soebarto V (2022) The thermal environment of housing and its implications for the health of older people in south Australia: a mixed-methods study. Atmosphere 13(1):96
29. Argyroudis SA, Mitoulis SA, Winter MG, Kaynia AM (2019) Fragility of transport assets exposed to multiple hazards: state-of-the-art review toward infrastructural resilience. Reliab Eng Syst Saf 191:106567
30. Markolf SA, Hoehne C, Fraser A, Chester MV, Underwood BS (2019) Transportation resilience to climate change and extreme weather events—beyond risk and robustness. Transp Policy 74:174–186
31. Fitzgibbons J (2019) Building inclusive resilience: exploring justice and social equity in urban resilience planning. Master's thesis, University of Waterloo
32. Greenwalt J, Raasakka N, Alverson K (2018) Building urban resilience to address urbanization and climate change. In: Resilience. Elsevier, New York, pp 151–164
33. Kumar S, Sharifi A (2022) SDGs in global south cities for building resilience to climate change. In: SDGs in the Asia and Pacific region. Springer, Cham, pp 1–29
34. Abuwaer N, Ullah S, Al-Ghamdi SG (2023) Building climate resilience through urban planning: strategies, challenges, and opportunities. Sustain Cities Changing Clim Enhancing Urban Resilience 2023:185–206
35. Antoszewski P, Krzyżaniak M, Świerk D (2022) The future of climate-resilient and climate-neutral city in the temperate climate zone. Int J Environ Res Public Health 19(7):4365
36. Cobbinah PB (2021) Urban resilience in climate change hotspot. Land Use Policy 100:104948
37. Friend R, Moench M (2013) What is the purpose of urban climate resilience? Implications for addressing poverty and vulnerability. Urban Clim 6:98–113

38. Cheshmehzangi A, Zuo J, Sharifi A, Zhang R, Ziafati Bafarasat A, Zhao J (eds) (2025) Designing healthy buildings and communities: shaping a climate-resilient future. Springer, Singapore
39. Cheshmehzangi A, He BJ, Sharifi A, Matzarakis A (2023) Climate change, cities, and the importance of cooling strategies, practices, and policies. In: Climate change and cooling cities. Springer, Singapore, pp 2–19
40. Bouzarovski S (2015) Retrofitting the city: residential flexibility, resilience and the built environment. Bloomsbury Publishing, London
41. Horgan D, Dimitrijević B (2020) Social innovation in the built environment: the challenges presented by the politics of space. Urban Sci 5(1):1
42. Schipper EL, Ayers J, Reid H, Huq S, Rahman A (2014) Community-based adaptation to climate change: scaling it up. Routledge, New York
43. Beceiro P, Brito RS, Galvão A (2022) Assessment of the contribution of nature-based solutions (NBS) to urban resilience: application to the case study of Porto. Ecol Eng 175:106489
44. Prabhakar AC (2024) Driving economic prosperity: fostering job-oriented sustainable and inclusive development in India. Open J Bus Manag 12(4):2854–2885
45. Cheshmehzangi A, Dawodu A, Sharifi A (2021) Sustainable urbanism in China. Routledge, New York
46. Dawodu A, Cheshmehzangi A, Sharifi A, Oladejo J (2022) Neighborhood sustainability assessment tools: research trends and forecast for the built environment. Sustain Fut 4:100064
47. Sharifi A, Dawodu A, Cheshmehzangi A (2021) Neighborhood sustainability assessment tools: a review of success factors. J Clean Prod 293:125912
48. Geerling L (2021) Rotterdam: the resilient delta city. In: Adaptation urbanism and resilient communities. Routledge, New York, pp 72–90
49. Gustafsson HR, Kelly EA (2016) Developing the sustainable city: Curitiba, Brazil, as a case study. In: How cities will save the world. Routledge, New York, pp 81–96
50. Widjaja G (2023) Economic development transformation with environmental vision: efforts to create sustainable and inclusive growth. Kurdish Stud 11(2):3154–3177
51. Ziafati Bafarasat A, Sharifi A (2024) How to achieve a healthy city: a scoping review with ten city examples. J Urban Health 101(1):120–140
52. Pieterse DE (2008) City futures: confronting the crisis of urban development. Bloomsbury Publishing, London
53. Pourroostaei Ardakani S, Cheshmehzangi A (2023) The role of big data analytics in urban systems: review and prospect for smart transport and healthcare systems. In: Big data analytics for smart transport and healthcare systems. Springer, Singapore, pp 1–26
54. International WELL Building Institute (2024) WELL community standard. Available from https://www.wellcertified.com/certification/community/

Heat-Health Vulnerability in Developing Metropolitan Areas: A Spatial Analysis in Tehran, Iran

Tandis Sulaimani and Mahdi Suleimany

Abstract Heat anomalies in metropolitan areas induced by global warming and urban heat islands (UHIs) constantly expose vulnerable citizens to health challenges, such as heat stroke. Moreover, unjust metropolitan policies have led to unequal adaptive capacities to cope with such challenges. Therefore, a comprehensive assessment of heat-health vulnerability (HHVA) in these areas is required to address the current deficiencies. The Tehran metropolitan area, as the capital city of Iran, is a case where citizens are unevenly exposed to urban heat and, in some parts, have inadequate access to essential facilities due to rapid population growth. Thus, we spatially analyse the heat-health vulnerability of the Tehran metropolitan area as a first step. We implement thermal remote sensing techniques on the Landsat 8 satellite data using a PyQGIS plugin to assess citizens' heat exposure. Additionally, we incorporate related socio-economic, environmental, and urban planning variables to analyse sensitivity and adaptive capacity. Afterward, we recommend urban cooling solutions to mitigate the studied vulnerability. The results indicate that Districts #03 and #01 are the least vulnerable as they are less exposed to heat, and also, their under-5-year-old population rate and poverty index are low. Moreover, the considerable area of urban greenery and proper access to cooling infrastructure have improved the adaptive capacity. In contrast, Districts #09 and #21 are considered the most vulnerable areas of the case due to higher heat exposure, high poverty index, and less access to critical infrastructure. The study provides a practical framework for ongoing heat-health vulnerability assessments to address the evolving challenges of urban heat and proposes urban planning strategies enhancing heat justice and resilience across the Tehran metropolitan area. The most crucial proposed solutions are to maintain and develop blue–green infrastructure by using climate-compatible species and reduce the intensity of urban heat stress by providing sufficient shading and open spaces. A just distribution of urban cooling centres is also highly recommended.

Keywords Heat-health vulnerability assessment · Spatial analysis · Tehran metropolitan area · Urban cooling solutions

T. Sulaimani · M. Suleimany (✉)
Urban Planning and Management, University of Tehran, Tehran, Iran
e-mail: mi.suleimany@gmail.com; mi.suleimany@ut.ac.ir

A. Cheshmehzangi et al. (eds.), *Designing Healthy Cities*, Urban Sustainability,
https://doi.org/10.1007/978-981-96-5538-0_2

1 Introduction

Although urban heat is often referred to as the invisible climate hazard [1], the Annual Disaster Statistical Review in 2017 ranked extreme heat as one of the most fatal natural disasters worldwide [2]. Extreme heat is also a leading cause of climate-induced disease and mortality [3, 4], especially in Asia [5]. Both climate change, due to the greenhouse gas emissions resulting from human activities, and urban heat islands (UHIs) contribute to intensifying urban heat and its adverse effects [1, 6]. As a result of the rapid city growth, UHIs are synergistically exacerbated by extreme heat events and intensify the urban heat side effects, particularly in metropolitan areas [7–9].

The adverse heat effects extend beyond heat-induced mortalities and include aggravating underlying diseases such as dyspnoea, cardiac distress, diabetes, and increased risk of preterm birth [10, 11]. Extreme heat conditions may lead to syncope, muscle cramps, organ failure, and even death by impacting the human body's natural temperature regulation mechanism and the circulatory and nervous systems [12, 13]. Its side effects may also be responsible for increasing the risk of suicide [14], growing hospital admissions for mental health problems [15], fatal road accidents [16], and the number of violent crimes observed [17]. In the COVID-19 pandemic era, reports indicate that heat-related diseases imposed significant pressure on medical systems and infrastructure worldwide as population groups vulnerable to heat stress and those at higher risk for severe COVID-19 largely overlapped [5, 18].

Since urban heat is a compound hazard, its adverse effects and citizens' heat vulnerability depend on both intrinsic (personal) and extrinsic (environmental) factors [19]. Therefore, heat-induced risks are distributed unequally from a spatial point of view in the cities, and how city planning and previous policies shaped the built environment plays a substantial role in these differences [1]. Consequently, the responsibility of improving heat justice is on the urban planners who, as a result of growing heat concerns for public health, should analyse spatial patterns of vulnerability and the areas most in need of future interventions, including mitigation strategies, adaptation policies, and effective heat management techniques [20]. Developing a Heat Vulnerability Index (HVI) is one of the main approaches to analyse cities' conditions in the global warming era and identify communities demanding more interventions. The heat vulnerability maps can assist urban policy-makers in prioritizing places for locating urban parks and green spaces, installing cooling centres, and providing urban facilities efficiently.

One of the common approaches for urban heat resilience planning is the integration of heat mitigation and adaptation strategies into city development. The first can reduce heat exposure through interventions such as land use regulations and urban design policies, and the second one maintains the coping capacity via planning for energy, health, and emergency responses [1]. Multiple studies have investigated mitigation and adaptation strategies and analysed their effectiveness under different conditions. These strategies mainly consist of increasing the surface albedo by modifying the materials and colours of urban buildings, roofs, and walls [19], designing cool pavements [21], optimizing air flow through wind corridors [22], augmenting ambient shading [23], incorporating urban blue–green infrastructure [24], protecting and increasing urban forestry, and water features and bodies [25, 26], developing city cooling centres [27], and implementing heat health warning systems [28].

Although urban heat and its related topics are studied in a considerable body of literature, analysing cities' vulnerability to heat and its side effects on public health is highly case-specific, as diverse communities have different environmental and socio-economic characteristics. Hence, this study aims to analyse heat vulnerability and its spatial patterns in the Tehran metropolitan area, Iran's capital and its most populous city. Given its population of over 9 million in the city and approximately 14 million in the broader region, it presents a compelling case for carrying out a heat-health vulnerability assessment [29]. Rapid urbanization and climate change have exacerbated heat exposure risks, necessitating tailored interventions. Recent studies indicate an alarming increase in average temperatures in Tehran, with some projections implying that extremely hot days will grow dramatically in the future decades, posing significant health concerns to the metropolitan population [30, 31].

Accelerated population growth in Tehran, along with the absence of proper urban planning and climate management, has brought about considerable challenges and made the city more prone to heat vulnerability. [32] identify migration as the most challenging and critical driver of urban problems in this metropolis. This population growth has increased the demand for urban services and infrastructure, which has tripled the city territory between 1988 and 2014 [33] and imposed significant changes on land cover patterns and the city's natural resources. Furthermore, the mentioned trends have exacerbated the frequency and intensity of UHIs within this city [34], posing crucial threats to citizens' climatic comfort and public health. In addition to posing a threat to human health, the urban heat island effect also has an impact on energy consumption and air quality, which further increases public health risks [35, 36]. Furthermore, the socio-economic disparities within Tehran's diverse population play a vital role in shaping vulnerability to heat-related health outcomes, as underprivileged communities, often residing in poorly equipped neighborhoods, are at greater risk due to the low availability of healthcare and cooling facilities. Identifying these social determinants of health becomes crucial for allocating resources and developing effective adaptation and mitigation strategies as urban heat levels rise.

Moreover, Tehran's distinctive geographical features, situated at the foothills of the Alborz mountains, contribute to diverse microclimates that alter heat distribution patterns, highlighting the necessity for localized evaluations [30]. Such conditions warrant an urgent need for targeted assessments that can inform public health policies and urban planning practices. By addressing the intricate interactions between climate change, urbanization, and public health in Tehran, this study aims to not only fill a critical gap in the literature but also offer actionable insights for urban planners and policymakers. These findings will aid in the development of strategies to improve community resilience to extreme heat, ultimately benefiting public health and sustainable urban living. Consequently, the current study seeks to answer two research questions:

1. What are the indicators of urban heat vulnerability for metropolitan areas?
2. What is the spatial pattern of heat vulnerability in the Tehran metropolitan area, based on its 22 municipal districts?

Responding to these questions, this study first develops a heat vulnerability framework for the Tehran metropolitan area based on the Urban Heat-Health Vulnerability approach, including exposure, sensitivity, and adaptive indicators. Afterward, it analyses the spatial patterns of heat-health vulnerability within Tehran's metropolis to prioritize its districts for cooling interventions. Ultimately, it recommends some strategies to improve the heat resilience of identified vulnerable areas which can assist policy-makers in designing more efficient related policies.

2 Urban Heat-Health Vulnerability

Although the concept of vulnerability is central to climate change research, its definition remains multifaceted and context-dependent. The debate surrounding vulnerability's nature—whether it is intrinsic to systems or contingent on external stressors—further complicates its assessment, yet the growing significance of climate change requires a robust framework for vulnerability assessment, especially when it comes to climate-sensitive diseases [37]. Vulnerability assessments continue to evolve as climate-related challenges get more complicated, indicating a continuous transition from science-driven to policy-driven approaches [37].

The assessment of climate change vulnerability has undergone substantial evolution in response to changing stakeholder needs and a variety of contexts [37]. There are four key aspects that are covered by various vulnerability models and assessment techniques shaping the nature of the assessments: the specific subject of the assessment (e.g., extreme heat), the spatial scope (geographic units such as cities), the temporal scope (current or future vulnerabilities), and the selection of qualitative and quantitative methods and data for analysis [5]. Researchers can better address the implications of climate change and develop effective adaptation strategies by precisely defining the context and type of vulnerability.

Three predominant models emerge in this discourse, each offering a distinct perspective [37]. The risk-hazard framework views vulnerability as a function of external hazards, providing insight into how environmental changes might adversely influence communities and ecosystems by highlighting the relationship between external climate hazards and their potential effects on systems [38]. In contrast, the social constructivist framework outlines the role of socioeconomic variables and emphasizes that vulnerability differs greatly between social groups, rather than being evenly distributed throughout populations [39]. Finally, the integrated framework, as articulated by the IPCC, synthesizes features of both previous models by proposing that vulnerability arises from a system's exposure to climate stressors alongside its inherent sensitivity and adaptive capacity [40]

An integrated approach is necessary for a holistic assessment that captures the interplay between both environmental variables and socio-economic conditions—both external and internal factors. To date, no standardized reference system has been defined for heat vulnerability assessment; however, the indexing method is the most commonly used modelling technique [41]. In this regard, Heat-Health vulnerability assessment (HHVA) is an approach developed based on the population vulnerability framework, in which vulnerability is a function of exposure, sensitivity, and adaptive capacity as shown in the following equation [42]:

$$Heat - Health\ Vulnerability = (Exposure + Sensitivity) - (Adaptive\ Capacity)$$

As shown in the equation, while the growth in exposure and sensitivity rates leads to the exacerbation of vulnerability, the adaptive capacity has the opposite relation [43].

- ***Exposure rate*** refers to the frequency, intensity, and [spatial] distribution of a natural or anthropogenic hazard and its consequences [44]. The most important indicators representing this component are the air temperature and near-surface temperature obtained by stationary weather stations or calculated using satellite images and thermal remote sensing. Air pollution or relative humidity may also indirectly affect the heat exposure rate, especially in metropolitan areas [5].
- ***Sensitivity*** refers to the degree to which a system may be affected by hazards [40] and is reflected by demographic, socioeconomic, and health indicators. Li et al. [41] categorized the most frequent heat sensitivity variables into age, financial condition, education, population density, race/immigration, and underlying disease status through a comprehensive review of more than 70 related studies [41]. Numerous studies consider over-65-year-old and under-5-year-old inhabitants as sensitivity indicators as they are more susceptible to heat-related health consequences than other age groups [45]. Economic indicators such as low-income households [46], unemployment rate [47], and housing conditions [48] are also widely used due to their impact on access to fundamental facilities [45]. Multiple researchers have included underlying diseases and the disability rate in their studies, considering their effect on individual physical coping capacity [49, 50].
- ***Adaptive capacity*** is also defined as the ability of a system to reduce the side effects or consequences of hazards [28]. This component mainly consists of indicators demonstrating population's access to fundamental infrastructures that are effective in reducing heat exposure and sensitivity, such as residential facilities and cooling centres [5].

It is necessary to mention that there is no solid consensus regarding the classification of sensitivity indicators and adaptive capacity [5]. Also, there are no criteria for selecting indicators, and this relies mainly on the local contexts [41]. Table 1 represents the indicators opted for this study based on the contextual conditions and available data regarding exposure, sensitivity, and adaptive capacity.

Table 1 Heat-health vulnerability indicators used in this study based on the available data

Indicator	Variable	Description/measurement	References
Exposure rate	LST	6-Year average of maximum land surface temperature from 2016 to 2021	[4, 44, 51, 52]
Sensitivity rate	Older adults	Percentage of the over-65-year-old residing population	[9, 28, 45, 53]
	Children population	Percentage of the less-than-5-year-old residing population	[41, 44, 54, 55]
	People with disabilities	Percentage of residing population with a disability	[5, 51, 41, 55]
	Urban poverty	A combination of four sub-variables: the percentage of residential units with an area of less than 50 square meters; the number of households in a residential unit; the percentage of the total working population; the average price of one square meter of a residential unit	[5, 50, 54, 56, 57]
Adaptive capacity	Access to HVAC systems	Percentage of households with access to air conditioner	[7, 9, 43]
	Access to cooling centres	Distance from the census tract to the nearest cooling center	[55, 58– 60]
	Green space	Percentage of green spaces and park areas	[7, 8, 52]

3 Materials and Methods

3.1 Case Study: The Tehran Metropolitan Area, Iran

The Tehran metropolitan area is the capital and most populous city of Iran, with over 730 square kilometers of area and a population of approximately 8.7 million people [29]. This metropolis is located in the north of the central Iranian plateau, with a latitude of 35° 30′–35° 51′ north and a longitude of 51°–51° 40′ east [61]. Tehran is developed on the southern slope of the Alborz mountains and the northern edge of the central desert of Iran in a relatively flat plain, which has led to the difference in the climate of its northern and southern districts. This metropolitan area has 22 municipal districts with an average temperature difference of 3 °C [30]. Tehran, with its mostly hot and dry climate (except for the northern districts, which is cold and dry [30]) and low average annual rainfall, is not exempt from the global warming trend, and its urban climate has become continuously warmer in recent decades. On the other hand, the available official data indicates that the concentration of important national socio-economic centers in this metropolis has made it the most crucial population attraction center of Iran, exposing more citizens to heat [32, 62]. Figure 1 illustrates the location of the Tehran metropolitan area and its 22 municipal districts.

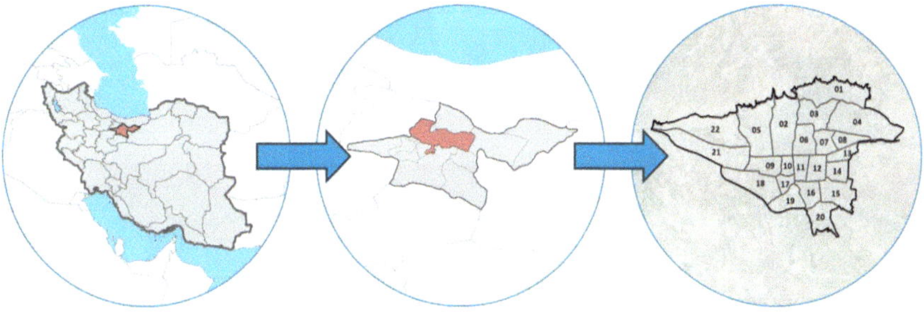

Fig. 1 The location of the Tehran metropolitan area and its 22 municipal districts

Table 2 Data sources and types used in this research

Data source	Data type	Data preparation
Satellite images	Mapped average of the highest annual land surface temperature (LST) in 2016–2021 period (°C)	Single channel algorithm on landsat satellite images using a PyQGIS plugin
National census	Sensitive age distribution	Mapping data with layer calculator tool and layer overlaying technique in QGIS application
	Disability distribution	
	Urban poverty distribution	
Municipal databases	Access to HVAC	Mapping data with layer calculator tool and layer overlaying technique in QGIS Application
	Access to cooling centres	
	Green space distribution	

3.2 Data Collection and Preparation

This study incorporates various sets of spatial data to analyse the Heat-Health vulnerability of the Tehran metropolitan area by developing a vulnerability index. These sets of data include climatic/environmental data (i.e., LST) retrieved through remote sensing techniques, socio-economic information (i.e., age, disability, and poverty distribution) based on national census data, and information about urban infrastructure and access to services (i.e., access to HVAC systems, access to cooling centers, and green space area) from municipal databases. Table 2 provides detailed insight into data sources and data collection techniques used in this research.

This research utilizes LST maps as a variable for the heat exposure indicator in Heat-Health vulnerability assessment. First, the metadata of the available Landsat satellite images for the hottest annual record in the six years from 2016 to 2021 were obtained from the USGS database[1] based on pre-defined criteria, including less than 2% cloud cover. Then, a Python QGIS plugin (PyQGIS) and Single Channel Algorithm (SCA) were implemented to extract LST maps from the Landsat 8 satellite

[1] https://earthexplorer.usgs.gov/.

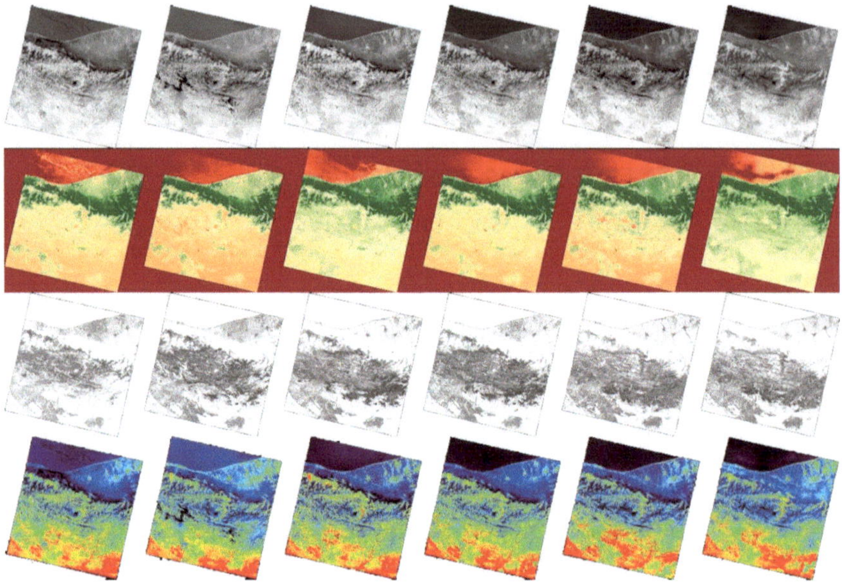

Fig. 2 The pictures are brightness temperature (1st Row), NDVI (2nd Row), land surface emissivity (3rd Row), and LST calculated using the single-channel algorithm (4th Row) for six selected satellite images (left to right from 2016 to 2021)

images. The PyQGIS was developed via Python programming to facilitate the process of generating LST maps from Landsat 5 TM, Landsat 7 ETM+, and Landsat 8 TIRS thermal infrared images [63].

Calculating the LST through the single-channel algorithm, which is one of the widely used techniques in this regard, requires several functional steps, including converting digital numbers (DN) to at-sensor radiance, converting spectral radiation to brightness temperature, calculating normalized difference vegetation index (NDVI) and land surface emissivity (LSE) ratio [64, 65]. These items are prepared as a part of processing LST maps and are provided in Fig. 2. The single-channel algorithm also demands the estimation of atmospheric water vapor during the satellite transit time, which is determined for the case study on the specified days using the World Meteorological Organization (WMO) index for the synoptic stations of Tehran metropolis [66].

3.3 Data Analysis

After preparing all the required data corresponding to the defined variables in Table 1, all the layers were standardized between 0.0 and 1.0 using the following equation:

Table 3 Value categorization based on standard deviation, developed by the authors, and adapted from Reid et al. [60]

Value	Category
≥ 2 SD below mean	Very low (vulnerability/exposure/sensitivity/adaptive capacity)
1–2 SD below mean	Low (vulnerability/exposure/sensitivity/adaptive capacity)
< 1 SD below mean	Low normal (vulnerability/exposure/sensitivity/adaptive capacity)
< 1 SD above mean	High normal (vulnerability/exposure/sensitivity/adaptive capacity)
1–2 SD above mean	High (vulnerability/exposure/sensitivity/adaptive capacity)
≥ 2 SD above mean	Very high (vulnerability/exposure/sensitivity/adaptive capacity)

$$Standardized\ X = (X - X_{min})/(X_{max} - X_{min})$$

These standardized non-weighted layers were first combined to develop exposure, sensitivity, and adaptive capacity indicators separately, and ultimately, overlayered to calculate the Heat-Health vulnerability index defined in the previous section. The analysis unit in this process was the 22 urban districts within the Tehran metropolitan area, and all the variables were assigned to these units using the zonal statistics tool in QGIS software. Ultimately, the mentioned districts were categorized into 6 levels based on their vulnerability and other indicators (i.e., exposure, sensitivity, and adaptive capacity), and the findings were interpreted based on the classification provided in Table 3.

4 Findings and Discussion

4.1 Exposure Rate

Figure 3 illustrates the heat exposure rate in the Tehran metropolitan area and its 22 districts. This map, as explained, is designed based on the LST data retrieved from the Landsat satellite images. The 6-year average of the hottest summer-time LST map shows that surface temperature variation can be intensified by about 25.6 °C in the Tehran Metropolitan area, leading to some urban points being over 50 °C hot in the summer. The results of the spatial analysis highlight the urgent need for targeted measures in the Tehran metropolitan area, where different areas experience varying degrees of excessive heat. According to the map, Districts #21 and #09 are the most exposed, and Districts #01 and #03 are the least exposed urban areas to heat.

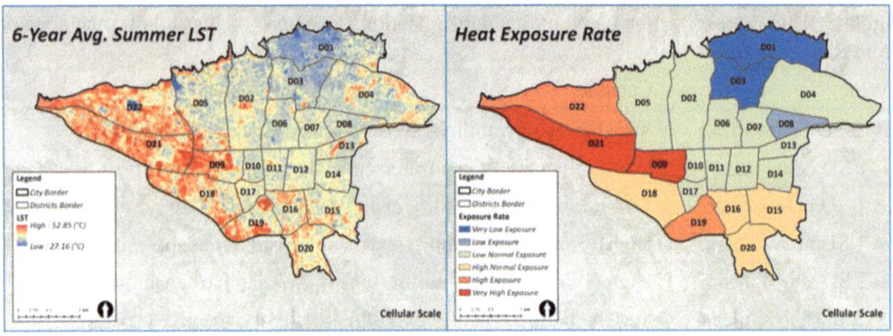

Fig. 3 LST (left) and heat exposure (right) maps of the Tehran metropolitan area

4.2 Sensitivity Rate

This situation is exacerbated for vulnerable populations in several districts, where sensitivity indicators reveal significant concentrations of individuals with underlying health conditions, disabilities, and low socioeconomic status. Figure 4 demonstrates the sensitivity map of the Tehran metropolis and its sub-indicators (i.e., age, disability, and poverty). The sensitivity indicator considers that citizens and different socio-economic strata have unequal ability to cope with heat, given their health (i.e., physical and mental well-being) and financial status. Sensitive groups, especially citizens with underlying diseases, disabilities, and low-quality residing conditions, should be prioritized in urban cooling interventions and required infrastructure provision in the heat mitigation and adaptation plans to improve urban climate (/heat) justice [67]. Districts #17 and #15 of the Tehran metropolitan area show the highest sensitivity rates in this case, as they contain a high proportion of sensitive age groups, people with disabilities, and low-income households. Conversely, Districts #5 and #2 are the least sensitive areas in the Tehran metropolitan area, considering their residents' demographic, health, and financial status. The findings underline how important it is to give these populations the highest priority when developing urban cooling strategies and infrastructure enhancements in order to ensure urban climate justice.

4.3 Adaptive Capacity

The analysis of adaptive capacity reveals stark contrasts across districts, as presented in Fig. 5. The aforementioned map provides insight into the adaptive capacity of the Tehran metropolitan area and its 22 districts, calculated based on three sub-indicators of access to HVAC systems, access to cooling centers, and percentage of green space area. Since urban districts within the Tehran metropolis are not equally developed, some areas have improper access to facilities and infrastructure necessary for coping with extreme heat events [57]. Besides, some households may not have access to

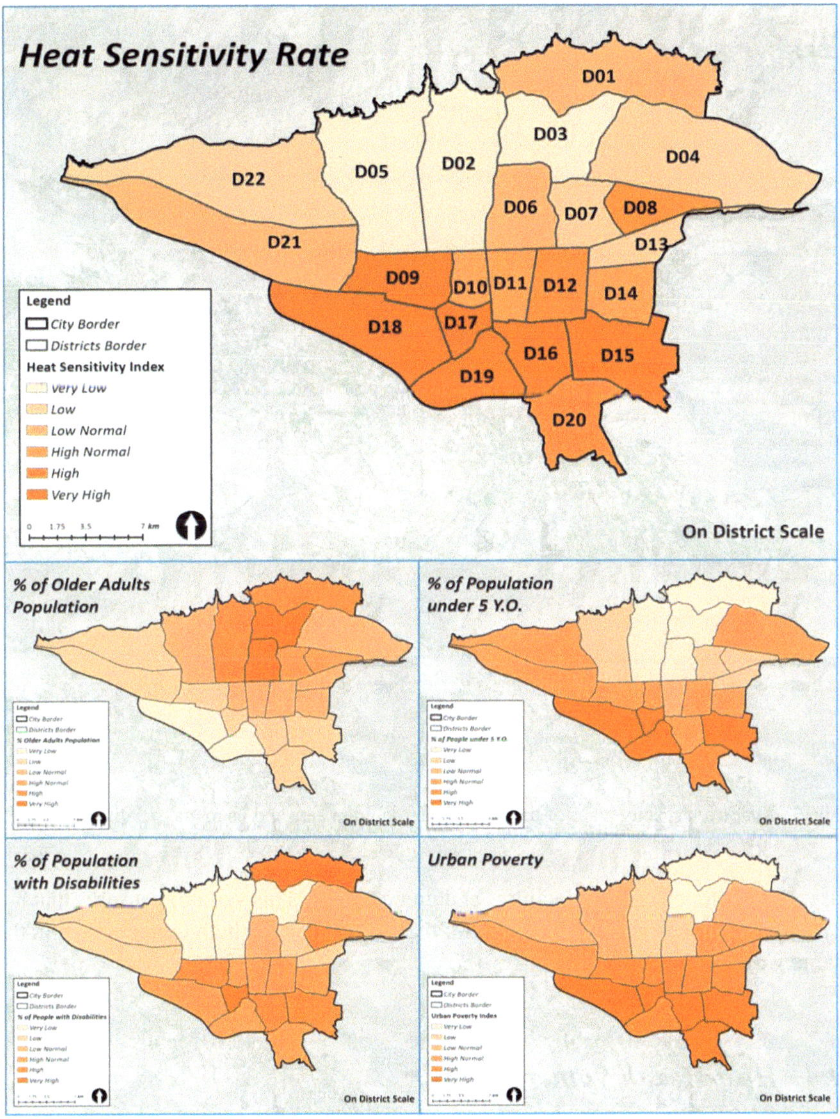

Fig. 4 Sensitivity maps of the Tehran metropolitan area and its municipal districts

suitable HVAC systems and basic infrastructure, such as electricity, on the building level, exacerbating their vulnerability to heat. Districts #21 and #22 of the Tehran metropolitan area have the least adaptive capacity, as they are newly added parts to the city and are almost under-developed. These areas do not have proper access to cooling centres and urban greenery. On the other side, Districts #01 and #04 demonstrate a robust adaptive capacity to heat as they have a considerable number

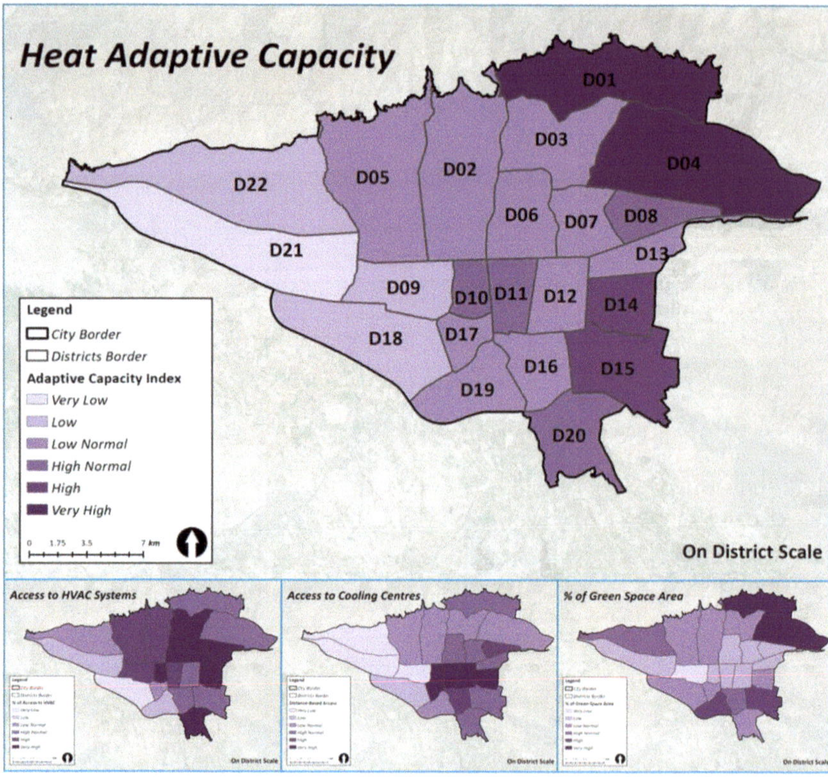

Fig. 5 Adaptive capacity maps of the Tehran metropolitan area and its municipal districts

of well-developed cooling facilities and green spaces. This variation ensures that not every resident will be able to cope with the implications of high temperatures in the same way.

4.4 Heat-Health Vulnerability

Figure 6 represents the heat-health vulnerability of the Tehran metropolitan area and its 22 districts. This map is the result of spatial overlaying of exposure, sensitivity, and adaptive capacity layers, integrating standardized climatic, health, and socio-economic variables. As shown in the map, Districts #09 and #21 of the Tehran metropolitan area are the most heat-vulnerable parts of the city and require the most interventions in climate and development plans to protect their citizens' health and well-being. In contrast, Districts #03 and #01 are the least vulnerable areas within the Tehran metropolis, considering the calculated Heat-Health vulnerability indicator. Analysing the spatial patterns of vulnerability indicates that the most vulnerable municipal districts are concentrated in the western and southern parts of the Tehran metropolitan area, calling for immediate action in climate and urban planning policies, whereas the northern and eastern parts are highly heat resilient. This spatial disparity stems from the unequal distribution of urban cooling centers, urban greenery, and wealth across the Tehran metropolis, indicating the crucial role of proper spatial and city planning in urban climate justice and heat resilience.

The findings of this study into heat-health vulnerability in the Tehran metropolitan area underline the necessity for ongoing vulnerability assessments as opposed to one-time analyses. Vulnerability to extreme heat is not static; socioeconomic changes, demographic shifts, climatic circumstances, urbanization, and infrastructure development all contribute to the alternation of [68, 69]. Tehran's rapid urbanization has resulted in notable changes to land use and coverage, exacerbating the health risks associated with the urban heat island effect [33]. Regular monitoring and data collection, incorporating real-time environmental changes and health outcomes, are essential for identifying emerging vulnerabilities and better prioritizing strategies.

According to projections, future climatic scenarios will most likely affect heat exposure and sensitivity rate. Tehran is expected to endure increasing temperatures and more frequent heatwaves owing to climate change [30, 31]. The elderly and other vulnerable groups, as well as those from lower-income backgrounds, are anticipated to be disproportionately affected, increasing the health risks associated with

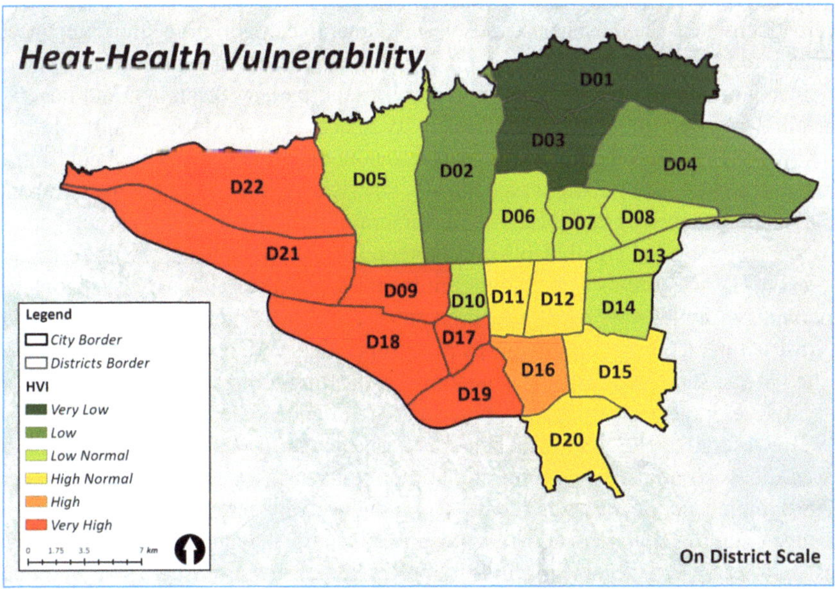

Fig. 6 Heat-health vulnerability of the Tehran metropolitan area and its 22 districts

excessive heat [69]. To lessen these effects in Tehran, it is crucial to implement effective strategies informed by continuous heat-health vulnerability evaluations. These enable prompt modifications to policy measures, guaranteeing that the city is able to successfully address the evolving challenges brought about by urbanization and climate change.

5 Conclusion and Recommendation

This research is conducted to spatially analyse the Heat-Health vulnerability of the Tehran metropolitan area and its 22 municipal districts. This study incorporates climatic, demographic, socio-economic, and urban planning variables to calculate the exposure, sensitivity, and adaptive capacity to assess the Heat-Health vulnerability of the case. The required data are collected and prepared through remote sensing and mapping techniques from different sources, including Landsat satellite images, national census, and municipal databases. Afterward, all the variables are combined via the spatial overlaying method in QGIS software, resulting in the Heat-Health vulnerability map of the Tehran metropolis.

The study sheds light on several novel and critical findings regarding the intricate interplay between urban heat exposure and public health vulnerabilities in Tehran. The incorporation of remote sensing techniques and spatial analysis facilitated an extensive mapping of Land Surface Temperatures (LST), unveiling concerning temperature levels above 50 °C in certain regions. Notably, Districts #09 and #21 were identified as the most heat-exposed and vulnerable, highlighting the urgent need for tailored heat mitigation and adaptation strategies. Furthermore, the analysis of socioeconomic and demographic indicators, including age, disability, and poverty, unveiled pronounced disparities in sensitivity to heat. The need for enhanced infrastructure and access to health resources was evident in districts with high concentrations of vulnerable populations, specifically Districts #17 and #15, demonstrating the role of socioeconomic variables in heat vulnerability. Moreover, the districts' adaptive capacities varied greatly from one another. Districts #21 and #22, with restricted access to HVAC systems and cooling centers, are prime illustrations of the challenges encountered by recently developed urban areas, which frequently lack the essential infrastructure. Conversely, greater adaptive capacities were observed in more established areas, such as Districts #01 and #04, indicating a direct correlation between urban development measures and community resilience to heat.

The results of the study underscore the necessity for continual vulnerability assessments to adjust to shifting climatic circumstances and urban development. The dynamic nature of vulnerability, as influenced by urbanization and climate change, requires frequent updates and proactive urban planning initiatives. This research serves as a valuable resource for policymakers, fostering them to incorporate thorough heat-health vulnerability assessments into urban development and health programs in Tehran to improve climate justice and public health resilience. As findings indicate the significant spatial disparity of the Heat-Health vulnerability across the Tehran metropolitan area—due to unequal distribution of sensitive groups, especially low-income households, along with the unfair provision of cooling facilities and urban blue–green infrastructure among municipal districts—this study recommends urban planning and design strategies to lower the Heat-Health vulnerability of the Tehran metropolitan area and improve its climate justice.

- ***Integrated Network of Urban Blue–Green Infrastructures:*** An integrated network of urban blue–green infrastructure (BGI) is considered an effective heat and flood mitigation strategy worldwide, which is practical in controlling urban heat islands and regulating microclimates, lowering the risk of urban floods, preserving water resources, alongside sequestering carbon, and is also beneficial for public physical and mental well-being [70, 71]. Considering the climatic condition of the Tehran metropolis, installing BGIs such as rainwater harvesting and bioswales would be useful. Implementing native and drought-resistant plant species in large-scale parks and urban forestry can be another option in this regard [72].
- ***Urban Cooling Centres:*** The fair distribution of urban cooling centers, which are public or semi-public places that people can use as shelter during extreme heat events, is considered one of the most crucial factors in communities' emergency response to heat [1]. Depending on the context, these centers are mainly established in schools, religious centers, health departments, libraries, malls, arcades, and the like [7, 73]. These centers should be located in the under-developed districts of the Tehran metropolitan area, especially where the heat exposure rate is higher.
- ***Natural Ventilation and Wind Corridors:*** The recent indiscriminate high-rise constructions in the Tehran metropolitan area have changed the microclimate by hindering the urban airflow, leading to the intensification of air pollution and UHI [74, 75]. Therefore, city planners and designers should facilitate airflow by maintaining the natural corridors in this metropolis, such as rivers and water bodies, along with the establishment of new air corridors on different scales, considering unblocked streets, orientation and width of streets, buildings' height and mass, the location of green open spaces, and conducting proper land use adjustment to increase the replacement rate of the stagnant hot air with cooler fresh air [76].
- ***High-Albedo Surfaces:*** Incorporating high-albedo surfaces, such as materials and colours that highly reflect solar radiation and have low heat absorption, into urban and building design standards can significantly reduce the ambient temperature [19]. In drought-prone cases like the Tehran metropolitan area, high albedo surfaces, such as cool roofs, would be considerably more efficient than water-intensive cooling solutions. Also, multimodal transportation should be integrated into urban street networks to support high-albedo pavements, especially in the over-exposed urban spaces [1].
- ***Shading:*** Shading by buildings, trees, and shading structures is one of the efficient heat mitigation strategies for cities, which improves thermal comfort by reducing surface temperature [23]. The suitability and efficiency of the shading methods

highly depend on the environmental conditions, the street network orientation, and the allocated budget. In cases where it is needed to reduce heat immediately, installing shading devices rather than planting shade trees is suggested [77]. Even interim design options, such as movable furniture and structures, are practical in this regard.

To efficiently implement recommended strategies for the case, they should be prioritized for each municipal district, based on the findings of this research. For instance, green infrastructure is highly demanded in District #09, while western districts, such as Districts #21 and #22, require more urban cooling centers. Although the results of this research are specific to the Tehran metropolis, the process can be utilized for other cases to spatially analyse their Heat-Health vulnerability, considering their contextual conditions. This research also faced some limitations, most of which were about the available data for the case. The lack of access to public health databases, such as health insurance, and general underlying disease data impeded us from comprehensively analysing the side effects of urban heat for public well-being in the Tehran metropolitan area. Future studies may collect these data through survey analysis for each municipal area and repeat this research with more specialized health variables along with the age and disability factors used in this paper.

Acknowledgements The authors would like to wholeheartedly thank Dr. Mojgan Taheri Tafti from the University of Tehran for her unwavering support and guidance.

References

1. Keith L, Meerow S (2022) PAS report 600: planning for urban heat resilience. APA Publication, London
2. Below R, Wallemacq P (2018) Annual disaster statistical review 2017. Centre for Research on the Epidemiology of Disasters (CRED)
3. Habeeb D, Vargo J, Stone B (2015) Rising heat wave trends in large US cities. Nat Hazards 76:1
4. Johnson DP, Stanforth A, Lulla V, Luber G (2012) Developing an applied extreme heat vulnerability index utilizing socioeconomic and environmental data. Appl Geogr 35:23–31
5. Szagri D, Nagy B, Szalay Z (2023) How can we predict where heatwaves will have an impact? A literature review on heat vulnerability indexes. Urban Clim 52:1
6. Almeida C, Furst L, Gonçalves A, Teodoro AC (2022) Remote sensing image-based analysis of the urban heat island effect in Bragança, Portugal. Environments 9:98
7. Bradford K, Abrahams L, Hegglin M, Klima K (2015) A heat vulnerability index and adaptation solutions for Pittsburgh, Pennsylvania. Environ Sci Technol 49:11303–11311
8. Maragno D, Dalla Fontana M, Musco F (2020) Mapping heat stress vulnerability and risk assessment at the neighbourhood scale to drive urban adaptation planning. Sustainability 12:1056
9. Wolf T, McGregor G (2013) The development of a heat wave vulnerability index for London, United Kingdom. Weather Clim Extremes 1:59–68
10. Barreca A, Schaller J (2020) The impact of high ambient temperatures on delivery timing and gestational lengths. Nat Clim Chang 10:77–82

11. Niu Y, Li Z, Gao Y, Liu X, Xu L, Vardoulakis S et al (2021) A systematic review of the development and validation of the heat vulnerability index. Curr Clim Change Rep 7:1
12. Barraclough KA, Blashki GA, Holt SG, Agar JWM (2017) Climate change and kidney disease—threats and opportunities. Kidney Int 92:526–530
13. Mora C, Counsell CWW, Bielecki CR, Louis LV (2017) Twenty-seven ways a heat wave can kill you: deadly heat in the era of climate change. Circ Cardiovasc Qual Outcomes 10:e004233
14. Florido Ngu F, Kelman I, Chambers J, Ayeb-Karlsson S (2021) Correlating heatwaves and relative humidity with suicide (fatal intentional self-harm). Sci Rep 11:22175
15. Trang PM, Rocklöv J, Giang KB, Kullgren G, Nilsson M (2016) Heatwaves and hospital admissions for mental disorders in northern Vietnam. PLoS ONE 11:e0155609
16. Wu CYH, Zaitchik BF, Gohlke JM (2018) Heat waves and fatal traffic crashes in the continental United States. Accid Anal Prev 119:195–201
17. Heo S, Choi HM, Lee J, Bell M (2024) A nationwide time-series analysis for short-term effects of ambient temperature on violent crime in South Korea. Sci Rep 14:3210
18. Bose-O'Reilly S, Daanen H, Deering K, Gerrett N, Huynen M et al (2021) COVID-19 and heat waves: new challenges for healthcare systems. Environ Res 2021:1
19. Fernandez Milan B, Creutzig F (2015) Reducing urban heat wave risk in the 21st century. Curr Opin Environ Sustain 14:221–231
20. Gamble J, Schmeltz M, Hurley B, Hseih J, Jette G, Wagner H (2018) Mapping the vulnerability of human health to extreme heat in the US. US Environmental Protection Agency, Washington, DC
21. U.S. Environmental Protection Agency (2012) Cool Pavements In: Reducing Urban Heat Islands: Compendium of Strategies. Draft. https://www.epa.gov/heat-islands/heat-island-compendium.
22. Xu Y, Wang W, Bingyin C, Chang M, Wang X (2021) Identification of ventilation corridors using backward trajectory simulations in Beijing. Sustain Cities Soc 70:102889
23. Lee I, Voogt JA, Gillespie TJ (2018) Analysis and comparison of shading strategies to increase human thermal comfort in urban areas. Atmosphere 9:91
24. Almaaitah T, Appleby M, Rosenblat H, Drake J, Joksimovic D (2021) The potential of blue–green infrastructure as a climate change adaptation strategy. Blue Green Syst 3:223–248
25. Middel A, Chhetri N, Quay R (2015) Urban forestry and cool roofs: assessment of heat mitigation strategies in Phoenix residential neighbourhoods. Urban For Urban Green 14:178–186
26. Yao L, Sailor DJ, Yang X, Xu G, Zhao L (2023) Are water bodies effective for urban heat mitigation? Evidence from field studies of urban lakes in two humid subtropical cities. Build Environ 245:110860
27. Meade R, Notley S, Akerman A, McCormick J, King KE et al (2023) Efficacy of cooling centres for mitigating physiological strain in older adults during daylong heat exposure. Environ Health Perspect 2023:1
28. Mallen E, Stone B, Lanza K (2019) A methodological assessment of extreme heat mortality modelling and heat vulnerability mapping in Dallas. Texas. Urban Clim 30:100528
29. Tehran Population by Administrative Divisions (up to Village Level)-Results of the 2016 National Population and Housing Census, *Statistical Center of Iran*, (2016). https://amar.org.ir/statistical-information/statid/52299
30. Bokaie M, Kheirkhah ZM, Daneshkar Arasteh P, Hosseini A (2016) Assessment of UHI based on the relationship between land surface temperature and land use in Tehran. Sustain Cities Soc 2016:1
31. Mansouri Daneshvar MR, Ebrahimi M, Nejadsoleymani H (2019) An overview of climate change in Iran: facts and statistics. Environ Syst Res 8:1
32. Khazaei K, Aliakbari I, Veicy H (2023) Forecasting the challenges of the spatial expansion of Tehran metropolitan area. Geogr Territor Spatial Arrang 13:131–160
33. Aghasi N, Estoque RC (2017) Tehran metropolitan area. In: Murayama Y, Kamusoko C, Yamashita A, Estoque RC (eds) Urban development in Asia and Africa. Springer, Singapore

34. SadeghiNia A, Alijani B, Zeaiean P (2013) Analysis of spatial–temporal structure of the urban heat island in Tehran. J Geography Environ Hazards 2013:1
35. Roshan G, Sarli R, Grab SW (2021) The case of Tehran's urban heat island, Iran: Impacts of urban 'lockdown' associated with the COVID-19 pandemic. Sustain Cities Soc 75:103263
36. Sedaghat A, Sharif M (2022) Mitigation of the impacts of heat islands on energy consumption in buildings: a case study of the city of Tehran, Iran. Sustain Cities Soc 76:1–15
37. Füssel HM, Klein R (2006) Climate change vulnerability assessments: an evolution of conceptual thinking. Clim Change 75:301–329
38. Dilley M, Boudreau TE (2001) Coming to terms with vulnerability: a critique of the food security definition. Food Policy 26:229–247
39. Adger WN, Kelly M (1999) Social vulnerability to climate change and the architecture of entitlements. Mitig Adapt Strat Glob Change 4:253–266
40. McCarthy JJ, Canziani OF, Leary NA, Dokken DJ, White KS (2001) Climate change 2001: impacts, adaptation, and vulnerability. In: Contribution of working group II to the 3rd assessment report of the IPCC
41. Li F, Yigitcanlar T, Nepal M, Thanh KN, Dur F (2022) Understanding urban heat vulnerability assessment methods: a PRISMA review. Energies 15:6998
42. Wilhelmi OV, Hayden MH (2010) Connecting people and place: a new framework for reducing urban vulnerability to extreme heat. Environ Res Lett 5:014021
43. Rathi SK, Chakraborty S, Mishra SK, Dutta A, Nanda L (2021) A heat vulnerability index: spatial patterns of exposure, sensitivity and adaptive capacity for urbanites of four cities of India. Int J Environ Res Public Health 19:283
44. Ahmad Kamal NI, Ashaari Z, Abdullah A, Mohd Kusin F, Yusuff F et al (2021) Extreme heat vulnerability assessment in tropical region: a case study in Malaysia. Clim Develop 2021:1
45. Nayak SG, Shrestha S, Kinney PL, Ross Z, Sheridan SC et al (2018) Development of a heat vulnerability index for New York State. Public Health 161:127–137
46. Hondula DM, Davis RE, Leisten MJ, Saha MV, Veazey LM, Wegner CR (2012) Fine-scale spatial variability of heat-related mortality in Philadelphia County, USA: a case-series analysis. Environ Health A Global Access Sci Source 11:16
47. Tran DN, Doan VQ, Nguyen VT, Khan A, Thai PK, Cunrui H, Chu C, Schak E, Phung D (2020) Spatial patterns of health vulnerability to heatwaves in Vietnam. Int J Biometeorol 64:863–872
48. Rinner C, Patychuk D, Bassil K, Nasr S, Gower S, Campbell M (2010) The role of maps in neighborhood-level heat vulnerability assessment for Toronto. Cartogr Geograph Inform Sci 2010:1
49. Loughnan M, Nicholls N, Tapper NJ (2012) Mapping heat health risks in urban areas. Int J Popul Res 2012:1
50. Qureshi AM, Rachid A (2022) Heat vulnerability index mapping: a case study of a medium-sized city (Amiens). Climate 10:113
51. Inostroza L, Palme M, de la Barrera F (2016) A heat vulnerability index: spatial patterns of exposure, sensitivity and adaptive capacity for Santiago de Chile. PLoS ONE 11:e0162464
52. Liu X, Yue W, Yang X, Hu K, Zhang W, Huang M (2020) Mapping urban heat vulnerability of extreme heat in Hangzhou via comparing two approaches. Complexity 2020:1–16
53. Conlon KC, Mallen E, Gronlund CJ, Berrocal VJ, Larsen L, O'Neill MS (2020) Mapping human vulnerability to extreme heat: a critical assessment of heat vulnerability indices created using principal components analysis. Environ Health Perspect 128:97001
54. Bao J, Li X, Yu C (2015) The construction and validation of the heat vulnerability index, a review. Int J Environ Res Public Health 12:7220–7234
55. Harris County Texas Public Health (2021) Vulnerability assessment indicators
56. Mitlin D (2001) Housing and urban poverty: a consideration of the criteria of affordability, diversity and inclusion. Hous Stud 16:509–522
57. Zebardast E, Ramezani R (2016) Evaluating urban poverty and its relationship with access to services in Qazvin city. J Fine Arts Arch Urban Plan 21:45–54
58. Harlan SL, Declet-Barreto JH, Stefanov WL, Petitti DB (2013) Neighbourhood effects on heat deaths: social and environmental predictors of vulnerability in Maricopa County. Environ Health Perspect 2013:1

59. Hayden MH, Wilhelmi OV, Banerjee D, Greasby T, Cavanaugh JL et al (2017) Adaptive capacity to extreme heat: results from a household survey in Houston. Weather, Climate, and Society, New York
60. Reid CE, O'Neill MS, Gronlund CJ, Brines SJ, Brown DG et al (2009) Mapping community determinants of heat vulnerability. Environ Health Perspect 117:1730–1736
61. Hosseini A, Porahmad A, Pajoohan M (2015) Assessment of institutions in sustainable urban management effects on sustainable development of Tehran. J Urban Plan Develop 2015:1
62. Hamdi K, Fathi S (2009) Effective factors in the expansion of immigration to Tehran metropolis. Urban Manag Stud 1:27–44
63. Isaya Ndossi M, Avdan U (2016) Application of open-source coding technologies in the production of land surface temperature (LST) maps from Landsat: a PyQGIS plugin. Remote Sens 8:413
64. Jiménez-Muñoz JC, Cristóbal J, Sobrino J, Sòria G, Ninyerola M et al (2009) Revision of the single-channel algorithm for land surface temperature retrieval from Landsat thermal-infrared data. IEEE Trans Geosci Remote Sens 47:339–349
65. Zhang J, Wang Y, Li Y (2006) A C++ program for retrieving land surface temperature from the data of Landsat TM/ETM+ band6. Comput Geosci 32:1796–1805
66. Liu L, Zhang Y (2011) Urban heat island analysis using the Landsat TM data and ASTER data: a case study in Hong Kong. Remote Sens 3:1535–1552
67. Suleimany M (2023) Urban climate justice in hot-arid regions: vulnerability assessment and spatial analysis of socio-economic and housing inequality in Isfahan, Iran. Urban Clim 51:101612
68. Girard LF, Nocca F (2020) Climate change and health impacts in urban areas: towards hybrid evaluation tools for new governance. Atmosphere 11:1344
69. Li F, Yigitcanlar T, Li W, Nepal M, Nguyen K, Dur F (2024) Understanding urban heat vulnerability: scientometric analysis of five decades of research. Urban Clim 56:102035
70. Mokhtarzadeh S, Suleimany M (2023) Who benefits more from urban cooling strategies? Exploring climate justice in vulnerable groups' access to blue–green infrastructure. In: Cheshmehzangi A, He B, Sharifi A, Matzarakis A (eds) Climate change and cooling cities. Springer, New York
71. Valente de Macedo LS, Barda Picavet ME, Puppim de Oliveira JA, Shih W-Y (2021) Urban green and blue infrastructure: a critical analysis of research on developing countries. J Clean Prod 2021:1
72. Sjöman H, Hirons AD, Bassuk NL (2015) Urban Forest resilience through tree selection-variation in drought tolerance in Acer. Urban For Urban Green 14:858–865
73. Nayak SG, Shrestha S, Sheridan SC, Hsu W-H, Muscatiello NA et al (2019) Accessibility of cooling centers to heat-vulnerable populations in New York State. J Transp Health 14:100563
74. Ranjbar-Saadatabadi A, Azadi M, Aliakbari-Bidokhti AA, Sadeghi-Hosseini AR (2005) Case study of Tehran heat island and its numerical simulation. J Earth Space Phys 31:63–78
75. Salehi E, Yavari A, Vakili F, Parivar P (2016) Assessing the impact of urban high-rise building on wind flow performance, case study: Tehran, District 22. J Urban Ecol Res 7:67–80
76. Wu K-L, Shan L (2024) Make way for the wind—promoting urban wind corridor planning by integrating RS, GIS, and CFD in urban planning and design to mitigate the heat island effect. Atmosphere 15:257
77. Qin S, Liu Y, Yu G, Li R (2023) Assessing the potential of integrated shading devices to mitigate overheating risk in university buildings in severe cold regions of China: a case study in Harbin. Energies 16:6259

Humanizing Al-Madinah Al-Munawarah Streets: From a Car-Centric City to a Healthy, Culturally Vibrant, and Climate-Resilient City

Alshimaa Aboelmakarem Farag and Rahma M. Doheim

Abstract The implementation of car-centric planning combined with the annual increase of visitors for Hajj and Umrah in Al-Madinah has led to a significant increase in car dependence on the streets, resulting in traffic congestion, air pollution, and a deterioration of public well-being. In response to Saudi Vision 2030, Al-Madinah is currently transitioning from a focus on cars to a focus on people in its planning. Al-Madinah's Municipality is seeking a significant transformation that will establish the city as a prominent example of urban development, prioritizing the well-being and quality of life of its residents. This chapter displays the case study of Al-Madinah as an example of a city taking significant action to humanize its streets, moving away from a focus on cars and instead creating spaces that are more livable, environmentally sustainable, and culturally vibrant. These efforts include enhancing the infrastructure to be more pedestrian-friendly, promoting the visual and cultural identity, improving accessibility to historical sites, increasing vegetation, enhancing diversity, enhancing public spaces, and improving the public transit infrastructure. There are ongoing challenges in attaining an appropriate balance between the use of cars and the needs of pedestrians. The chapter asserts that by giving priority to the needs of people and the local culture, cities can create vibrant spaces for people, a healthier and climate resilient environment.

Keywords Al-Madinah · Culture · Public spaces · Pedestrian-friendly · Healthy city · Streetscape · Quba avenue

A. A. Farag (✉)
Zagazig University, Zagzig, Egypt
e-mail: alshimaa_79@yahoo.com

R. M. Doheim
Al Yamamah University, Riyadh, Saudi Arabia

A. Cheshmehzangi et al. (eds.), *Designing Healthy Cities*, Urban Sustainability,
https://doi.org/10.1007/978-981-96-5538-0_3

1 Introduction

Al-Madinah Al-Munawarah, a Saudi Arabian city, has undergone significant urban expansion and has faced several challenges, notably related to transportation and the prevalence of car-centric urban planning [1]. The mobility system of Al-Madinah has primarily relied on private cars, with limited alternatives for public transport or active mobility available to both residents and visitors [2]. Moreover, it is common for households to own many cars due to cultural norms and the desire for privacy. This has resulted in a growing reliance on private cars, especially when women were granted permission to drive [3]. In addition, Al-Madinah receives millions of visitors every year for religious tourism associated with the Islamic calendar, including Hajj and Umrah [4, 5].

This increase in population has put escalating pressure on the existing transportation infrastructure, leading to congestion and inefficiencies in the transportation system. The city's road infrastructure is extensive, but it lacks the capacity to handle the growing need for cars on its roads. Hence, Al-Madinah people are suffering daily and continual issues due to the car-centric lifestyle. The traffic congestion poses a substantial problem, as commuters endure significant delays while suffering from compromised air quality caused by elevated levels of car emissions. In addition to its impact on the environment; the elevated amounts of emissions can worsen climate change, a global issue that needs attention [6]. The city's excessive dependence on cars has resulted in a deterioration of the people's quality of life, as they are compelled to allocate more time to their cars, whether it be for traveling to work or completing errands. This can result in heightened levels of stress and a decline in public well-being and quality of life [7]. In addition to health problems, the sedentary lifestyle resulting from excessive reliance on cars and a lack of physical activity contributes significantly to the elevated prevalence of obesity among residents [8]. Early findings from a study conducted by the WHO [9] indicate that a sedentary lifestyle is among the top ten causes of death and disability globally.

With the growing recognition of the negative impacts of car-centric planning on the environment and human well-being in Al-Madinah, the "Humanizing Cities" concept presents a promising alternative approach. The "Humanizing Cities" concept has gained significant attention, especially in the context of culture and urban development [10]. Al-Madinah has been transforming to humanize its public spaces, aiming to achieve vibrant public realms toward a healthy, vibrant, and resilient future. Since 2017, the "Humanizing Cities" program in Al-Madinah has been actively working to improve and enhance some neighborhoods and important streets. This study specifically examines the efforts to make the city's streets more human-friendly as a component of the entire program.

Humanizing cities is challenging, especially with the increase in urbanization and consequently the increase in car use. There is a lack of information concerning city humanization in different cultures, where the experience varies with the variation of cultures [11]. This chapter contributes to the knowledge of city planning by identifying the practices and learned lessons from studying the transformation of a car-centric city that faces many challenges into a humanized city.

This chapter sheds light on the problem of the long-term urbanization process that prioritizes cars, and its impact on health, quality of life, and climate resiliency in Al-Madinah City. The aim is to examine the capacity of the ongoing humanization initiatives to transform Al-Madinah away from car dependency toward a more healthy, culturally vibrant, and climate-resilient city. This chapter consists of five sections. The first section is the introduction of the study. The second section is the literature that explores the concept of humanization in the context of urban theories and the Saudi Vision as well as the concept of car-centric planning and the challenges that Al-Madinah City faces as a result of being car-centric. The third section is the methodology. The fourth section is the results that display the transformation of Al-Madinah into a culturally vibrant city, a healthier and climate-resilient environment. Lastly, the fifth section is a discussion of the success and impact of these humanization initiatives on improving the quality of life on the city's streets, as well as the challenges that arise.

2 Literature Review

2.1 Humanizing Cities Concept

Different urban theories provide valuable insights and ideas for the concept of "Humanizing Cities" and constructing more livable, inclusive, and sustainable cities by examining urban planning, design, and the human experience in cities. By prioritizing pedestrian-friendly design, cities can become more vibrant, secure, and environmentally responsible, and promote overall well-being. This approach, as Gehl [12] suggests, fosters a more socially sustainable urban environment, where public spaces become hubs for interaction, fostering a sense of inclusion in society. A city thrives when its residents are attracted to walking, cycling, and simply being present for longer periods in its public areas, which creates a sense of safety and security. A well-designed space with convenient walking distances attracts people, further enhancing safety and a sense of community. Sustainable practices are also improved when "green mobility" is prioritized. Walkable and bikeable access to public transportation strengthens its appeal, creating a harmonious connection between well-designed public spaces and an efficient transport system. Promoting walking and/or cycling as integral parts of daily life fosters a healthier city and a more sustainable future.

Lynch [13], introduces his idea of urban design explaining cities that are more human-centered in his theory known as "good city form." This theory emphasizes communities that can grow and evolve while yet maintaining a strong sense of identity and unity. Lynch argues that successful cities shouldn't be limited by rigid standards, but rather by flexible dimensions that cater to the unique needs and cultures of their residents. These dimensions, known as performance dimensions, represent general qualities essential for human well-being and cultural expression. The performance dimensions are the characteristics of the performance whose value varies with variations in culture. His five dimensions—vitality, sense, fit, access, and control—all contribute to creating a city that fosters a sense of community, belonging, and empowerment for its inhabitants. A well-designed city should be lively, identifiable, adaptable, and inclusive, and offer residents a sense of agency in their environment.

The diversity found in cities is a result of a vibrant mix of people with unique preferences, skills, desires, and resources living close together, which helps cities achieve a level of humanization that encourages vibrant streets, pedestrian activity, social interaction, and a sense of community. Jacobs [14], argued that cities become truly humanized through a careful balance of elements. Mixed-use areas, where people can live, work, and shop in close proximity, keep streets lively throughout the day. Short blocks with frequent intersections encourage walking, social interaction, accessibility, and urban vitality. A variety of building ages, from historic to modern, including affordable older options, adds visual interest and character. Finally, a moderate population density ensures enough people to support local businesses, and public services, and create a vibrant street life. Together, these elements humanize cities, making them vibrant, inclusive, and sustainable.

According to Yang et al. [10], there is no one-size-fits-all solution for creating human-centered cities, but one thing is certain: people's culture and intercultural understanding should be harnessed. Public cultural activities give cities identity and promote positive citizen interactions. Cultural diversity is promoted through the arts in public areas and festivals worldwide as part of city branding or urban renewal. They are vital for promoting interaction between different groups of people. Historic preservation is essential for city identity, pride, and a sense of belonging. Many historic settlements planned for walkability emphasize liveability at the threshold of private and public areas, where multifunctional flexible public spaces facilitate human interaction [15].

This exploration of urban theories highlights the multifaceted nature of the "Humanizing Cities" concept. Prioritizing pedestrian-friendly design, livability, vibrant public spaces, and diversity as advocated by [12, 14], fosters social interaction and a sense of community. Lynch [13] reinforces the importance of cultural sensitivity in urban design. Furthermore, historic preservation and public cultural activities are vital for fostering a city's identity and sense of belonging, and promoting positive citizen interactions, as discussed by Yang et al. [10]. Considering the cultural context and promoting intercultural understanding is crucial for creating human-centered cities that celebrate diversity, thrive in vibrant public spaces, and foster quality of life. Successful humanization requires a holistic approach that integrates diverse urban theories to create inclusive, sustainable, and culturally vibrant cities.

2.2 Al-Madinah: A Car-Centric City

The car-centric city is a term that refers to an urban environment where private cars are the primary mode of transportation, and infrastructure is primarily designed to accommodate and prioritize the use of cars over other modes of mobility [16, 17]. It is where walking, cycling, and public transport are dominated by car-centric approaches, thinking, decision-making, and policy design [18]. The dominance of cars in modern cities often results in pedestrian-unfriendly environments, and thus a range of negative implications for sustainable mobility, culture, public health, and quality of life [18–20].

The approach toward car dominance is particularly concerning in places like Saudi Arabia, where car-centric infrastructure already struggles to keep up with increasing demand. Based on a study conducted in Saudi Arabia in 2017, 65% of households own one car, whereas 35% own more than one car [21]. The number of licenses in 2019 was 2.4 times greater than the number of licenses in 2015 [21]. Extensive car-centric infrastructure struggles to cope with the ever-increasing number of cars, leading to a web of traffic congestion and air pollution [22]. Additionally, car dependence diminishes the quality of life, forcing residents to spend excessive time commuting and neglecting active lifestyles [23], compromising public health [24].

Al-Madinah Al-Munawarah, a Saudi city, see Map 1, holds significant cultural importance as it is one of the holiest cities in Islam and, thus receives millions of visitors every year for religious tourism associated with the Islamic calendar, including Hajj and Umrah. Additionally, Al-Madinah has encountered mobility difficulties, affecting the quality of life, particularly during the Hajj and Umrah seasons. The city represents an exemplary paradigm for a car-centric city that favors private cars over public transportation and non-motorized modes of mobility.

One of the primary reasons for the mobility issues in Al-Madinah is rapid urbanization and population growth. As the city continues to expand and attract more residents and visitors, the demand for transportation services has increased significantly [25]. Over the years, Madinah has witnessed dramatic increase in population density as a result of the fluctuating nature of religious tourism. Based on the 2011 census data, the population of Madinah was recorded as 1.3 million, with a growth rate of 2.9%. In the same year, the number of temporary residents, such as pilgrims and visitors, reached 9 million, making a noteworthy contribution to the overall population count. Al-Madinah is expected to experience substantial growth in the coming decades, with a predicted permanent residential population of 2.06 million and around 12 million annual visitors by 2040, which has added additional pressure on the current infrastructure [4, 5]. This increase in population has put escalating pressure on the existing transportation infrastructure, leading to congestion and inefficiencies in the transportation system.

Moreover, the public transportation system in Al-Madinah has faced several challenges that hinder its effectiveness in addressing the city's transportation needs [26]. Our frequent visits to Al-Madinah allow us to determine these challenges,

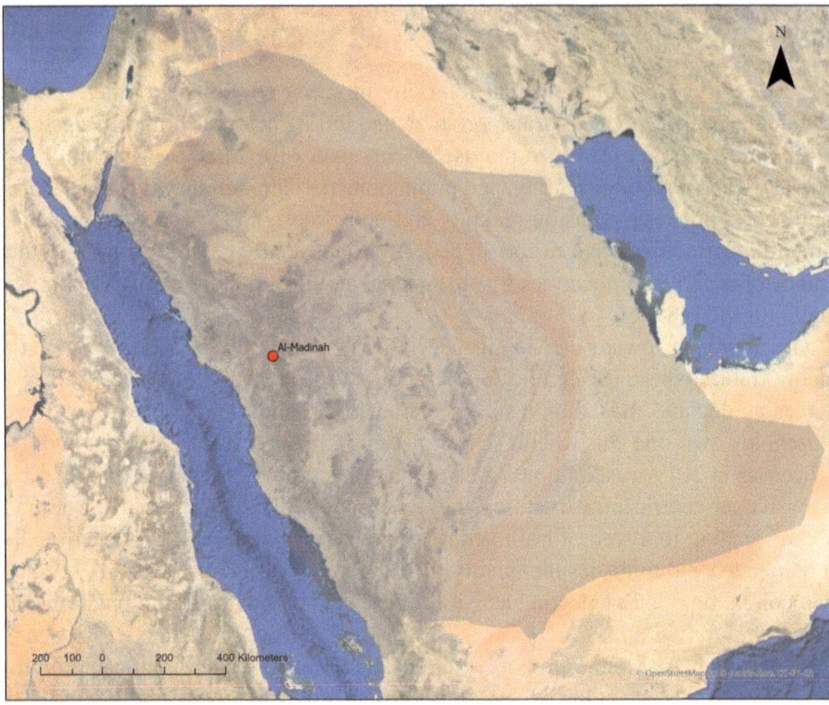

Map 1 AL-Madinah city is located in the western middle of Saudi Arabia. *Source* Authors (2024)

which include inadequate coverage, limited frequency of services, lack of integration between different modes of transport, and insufficient infrastructure to support sustainable mobility options. Urban mobility is likely to deteriorate in the absence of efficient public transportation, making private cars the most convenient mobility alternative [23]. Therefore, Aljoufie [27] emphasizes the need for effective public transportation planning in car-centric Saudi cities to address mobility issues.

2.3 Impacts of Car-Centric Planning

The car-centric planning of Al-Madinah has led to several challenges that impact the quality of life in the city, see Fig. 1. Some of how this planning approach undermines the quality of life of Al-Madinah's residents include:

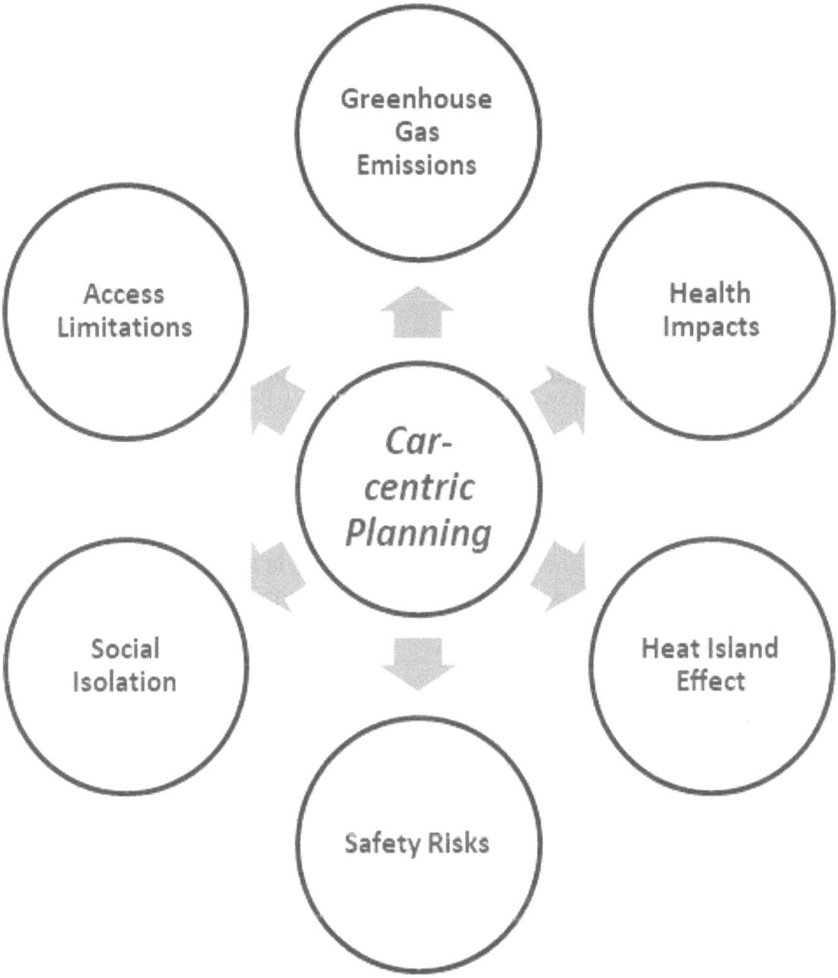

Fig. 1 Car-centric planning impact on the quality of life of Al-Madinah's residents. *Source* Authors (2024)

2.3.1 Greenhouse Gas Emissions

The reliance on private cars for transportation resulted in higher greenhouse gas emissions and increased levels of air pollution in Al-Madinah. The General Environmental Law and Rules for Implementation acknowledged that fuel combustion in urban areas significantly contributes to carbon monoxide, hydrocarbon, and nitrogen oxide emissions [28]. Moreover, the prioritization of cars over other modes of transport led to traffic congestion, especially during peak hours. Congested roads resulted in longer travel times, increased fuel combustion, and greenhouse gas emissions, all of which contribute to climate challenges [29].

2.3.2 Health Impacts

The increased air pollution resulting from heavy traffic has adverse effects on respiratory health and overall well-being [23]. Car emissions release harmful pollutants into the atmosphere, contributing to poor air quality and posing health risks to residents in Al-Madinah [30], leading to an increased risk of respiratory diseases and decreased well-being.

2.3.3 Heat Island Effect

Car-centric urban planning increases the urban heat impact. Common measures like green infrastructure are constrained by limited space due to the need for impermeable materials for car movement [31]. The extensive use of impermeable surfaces such as "Asphalt" in a car-centric city like Al-Madinah intensifies the urban heat island effect. This phenomenon leads to higher temperatures in urban areas compared to surrounding rural areas, further exacerbating heat-related issues [32]. Research shows that Al-Madinah is experiencing a significant increase in temperature that may lead to public health issues and potential risks associated with rapid climate change, and thus needs critical actions [33].

2.3.4 Safety Risks

Traffic congestion and inadequate pedestrian infrastructure can pose safety risks for residents, especially pedestrians and cyclists. Accidents and injuries related to road traffic may be more common in areas where car-centric planning prevails. According to statistics, Al-Madinah experienced a total of 257 deaths resulting from 14,595 road accidents in 2004. This number has risen to 569 fatalities caused by 43,543 road accidents in 2013. The fatality and accident rates have increased by 121% and 300%, respectively. Moreover, the rate of pedestrian casualties is on the rise due to factors such as overcrowding, an increased number of cars, reckless driving, and inadequate road safety rules [34].

2.3.5 Social Isolation

Empirical evidence indicates that cities with higher walkability, where people make more walking trips and less car travel, can lead to increased opportunities for social interaction and face-to-face social interactions within their community [35, 36]. People are susceptible to social isolation in car-centric cities such as Al-Medina, with a shortage of walking infrastructure and an increase in driving time. Research shows that Al-Madinah's arterial roads and intersections are overloaded, reducing average journey speeds during peak hours and increasing time spent on roads [37]. The long commutes limit individuals' access to community-based communication opportunities and their level of engagement in local residential communities [38], especially when the commute times are extreme, such as exceeding one hour per day [39].

2.3.6 Access Limitations

According to the Urban Prosperity Report issued by the Ministry of Municipal and Rural Affairs [4, 5], Al-Madinah's prosperity index for quality of life is reasonably high at 65.1%. However, it has average scores, particularly in terms of pedestrian access to public places and green areas. The extensive use of cars has led to allocating a privileged amount of urban space to cars, undermining the sense of community and the value of public spaces [40] crucial for a vibrant cultural life.

In addition, there is limited access to historical sites in Al-Madinah that can only be accessed by private cars and syndicate buses. This limitation is primarily caused by the city's mobility system, which consists of an extensive road network but needs more pedestrian connections.

To conclude, the over-reliance on cars has been shown to have multifaceted implications for the environment, quality of life, and well-being of people. The excessive reliance on cars in Al-Madinah has caused the city to become congested, with frequent traffic jams that pose risks to road users. It has also led to air pollution and health issues, as well as limited mobility for pedestrians. As a result, the quality of life in Al-Madinah has been negatively affected.

Therefore, there is a need for an effective solution such as "Humanizing City" to free urban areas from car dominance and prioritize human-centered spaces. By creating urban environments that encourage human interaction and pedestrian activities, cities can provide opportunities for people to interact, and thus lead to more vibrant, healthy, and resilient urban environments, ultimately reducing car dominance and promoting sustainable modes of transportation [41].

2.4 A Vision for Change: Humanizing Al-Madina's Public Spaces

The Kingdom of Saudi Arabia is exerting continuous efforts to enhance the livability of Saudi cities and make them human-centered to ensure high quality of life, promote well-being, preserve cultural identity and heritage, and foster social unity. This can be accomplished through a range of different programs and initiatives that are in line with Vision 2030. In 2018, the *Quality of Life Program* was launched, in which the development of urban areas to enhance livability is one of its objectives. Additionally, the *Vibrant Society*, one of the three pillars of the 2030 Saudi Vision, promotes Islamic, national, cultural, and heritage identities. Along with this, the Ministry of Municipal

Rural Affairs and Housing (MOMRAH) has launched the *Humanizing Cities initiative*. This initiative focuses on making cities more human-friendly, building public areas like plazas, pitches, parks, and green spaces, creating pedestrian paths and event spaces, enhancing city gateways, and placing attractive landmarks to improve resident well-being [42, 43].

The Saudi Vision 2030 is in accordance with urban theories that emphasize enhancing the quality of life in cities, incorporating the principles of livability, and promoting dynamic, vibrant, and sustainable communities. Inspired by the progressive goals of Saudi Vision 2030 and urban theories, Al-Madinah is pioneering the Humanizing Cities program.

The Humanizing Cities program in Al-Madinah began in early 2017, starting with the development and rehabilitation of the pilot neighborhood of "Hamra Al-Asad". By 2020, the program had targeted some neighborhoods for development and rehabilitation, such as some main streets in the city [44]. This study focuses on the initiative of humanizing the city's streets as part of the comprehensive program.

3 Methodology

This research adopts a comparative case study approach focusing on analyzing the transformation of Al-Madinah Al-Munawara from a car-centric city to a more human-centered urban environment through visual analysis. The study focuses on comparing specific streets within Al-Madinah before and after the transformation using literature, Google Earth imagery, photos, and site observation to assess the changes in urban design and infrastructure including land use, street layouts, green spaces, and pedestrian pathways. Additionally, the comparison allows us to evaluate the impact of this transformation on promoting pedestrian activity, reducing car use, community engagement, accessibility, and overall urban livability.

The study areas in Al-Madinah were chosen based on the identification of key streets that have undergone considerable transformation. These streets included areas surrounding major religious sites, and commercial zones. The selection criteria included the level of transformation and availability of data.

4 Results

The humanization efforts go beyond mere infrastructure changes to enhance the quality of life in Al-Madinah's streets and create a healthy and resilient environment. This includes creating car-free zones, developing public squares for events, and improving sidewalks and bike lanes. The city is also revitalizing historic buildings, supporting local artists, and hosting cultural festivals. To reduce car dependence, the city is expanding its bus network, and integrating different transportation alternatives.

Furthermore, the city is preserving the streets' cultural identity by using traditional architecture, incorporating public art, and increasing green spaces.

According to Al-Madinah Regional Development Authority [45], the project has focused on the following actions:

1. Developing the streets' urban design plans
2. Developing the residential building and commercial facades
3. Creating open spaces and designating parking lots for visitors
4. Increasing the vegetation
5. Creating streetscapes such as seating places, drinking coolers, and fountains

Below are the crucial endeavors of this project including Quba Avenue, Ohod Avenue, Al-Salam Avenue, and Hijra Road, see Map 2.

4.1 *Qeba Avenue*

Qeba Avenue, the major route connecting the Prophet's Mosque "Al-Masjid Al-Nabawi" to the Qeba Mosque, is a straight path measuring 3.6 km in length. It holds great significance in the history of Islam and Muslims due to its rich Islamic heritage and its association with the life of the Prophet Muhammad, peace be upon him, and his esteemed companions. It serves as a pathway for some of the Prophet's prayer halls and historical sites such as the Qeba Mosque, the first mosque built in Islam.

Qeba Avenue is one of the most significant and recent humanization endeavors to enhance the quality of life for residents and also for visitors who constitute a substantial section of the population in Al-Madinah City. This project aims to revive the Sunnah of the Prophet by facilitating pedestrian mobility between Al-Masjid Al-Nabawi and Qeba Mosque. To accomplish this, the congested, car-centric Qeba Road had to be converted into a secure pedestrian path.

The project endeavor has focused on upgrading the road network and infrastructure to improve transportation efficiency, facilitate pedestrian access to mosques, address congestion issues, and enhance the security and safety of visitors. Furthermore, there is a plan to restore and enhance the historical sites in the area, encompassing a total of 57 locations that include wells, farms, and orchards. The focal point of the project is the enhancement of the place's vibrancy and the improvement of the avenue's visual identity [45].

Frequent visits to the street and analyzing the visual data of Qeba Avenue, see Table 1 and Fig. 2, show the differences before and after humanization efforts and how the actions taken improve the street's livability and place vibrancy. There has been a significant change in the road that was previously dominated by cars and lacked green spaces, seating areas, and aesthetic elements that would attract people to linger and interact. The narrow sidewalks were a challenge for pedestrians, impeding their ability to stroll safely and securely. If the built environment was made more human-friendly, it would lead to a decrease in noise, pollution, and congestion, and an improvement in the quality of life for both residents and visitors.

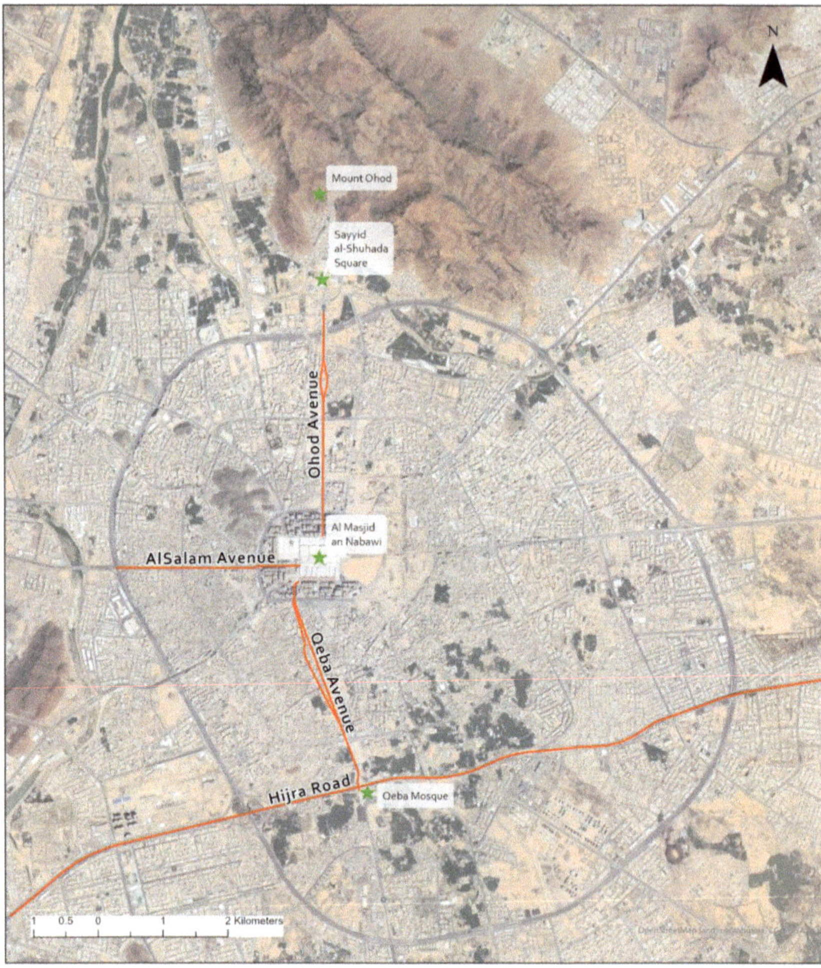

Map 2 Qeba Avenue, Ohod Avenue, Al-Salam Avenue and Hijra Road. *Source* Authors (2024)

The renovation process involved prioritizing pedestrian zones, widening sidewalks, boosting plant and streetscape components, and enhancing the visual identity by incorporating public art installations like calligraphy on walls and signs, and geometric patterns on walls and pavement, which are inspired by Medina's built and natural environment as well as its cultural characteristics. The introduction of public art installations added cultural value to the street, making it a destination for art enthusiasts and tourists. The improved aesthetics, such as landscaping and lighting design, created a visually appealing setting that attracted people to spend time outdoors. It effectively enhanced the place's visual identity while conserving its cultural heritage and ancient structures during the development process [46]. According to Mughairy

Table 1 The difference between the car-centric street in 2016 and the renovated Qeba Avenue in 2023

Qeba avenue		Before (2016)	After (2023)
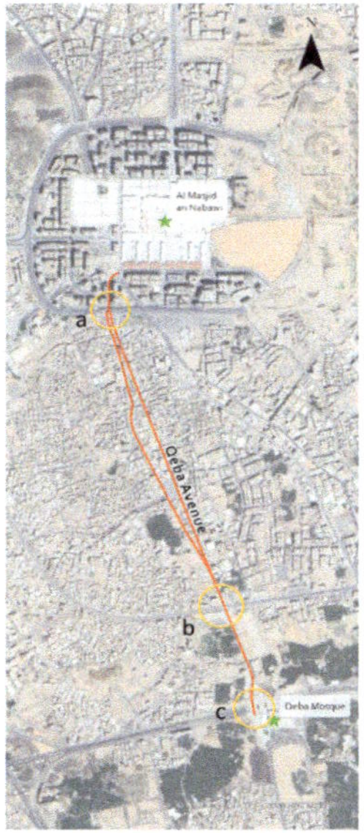	(a)		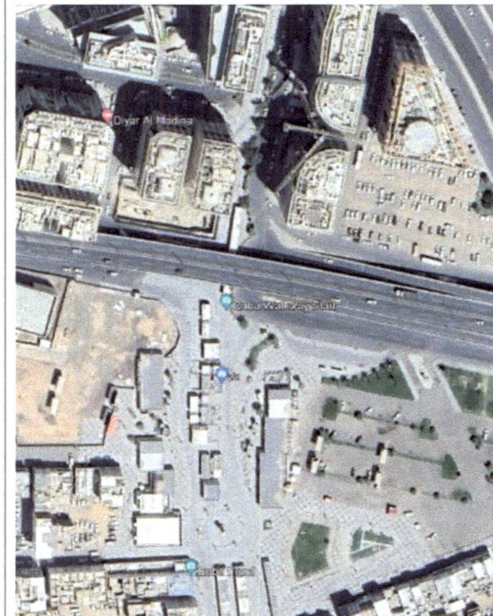

(continued)

Table 1 (continued)

Qeba avenue		Before (2016)	After (2023)
	(b)		

(continued)

Table 1 (continued)

Qeba avenue		Before (2016)	After (2023)
	(c)		

Fig. 2 The difference between the car-centric street (on the left) and the renovated Qeba Avenue (on the right). *Source* [45]

et al. [44], Qeba Avenue was renovated with careful consideration of the religious and cultural importance of the place.

The humanization of Qeba Avenue also enhanced the place's vibrancy by creating a more attractive and engaging public realm. One of the significant improvements is the diverse social activities that have taken place in the avenue. These activities include events, strolling, and making purchases in stores where visitors can find an abundance of local products, gifts, and souvenirs. Additionally, there are designated seating areas that provide people with the opportunity to observe pedestrians and various activities. There are also designated spots for children to engage in play, as well as establishments such as restaurants and cafes. Furthermore, visitors can find landmarks that are designed with consideration to the place's identity that offer the potential for creating lasting memories. Events and activities could now be hosted more effectively along the avenue, further activating the space and fostering community engagement.

Ultimately, this project has contributed to the revival of the religious tradition of walking between the Qeba Mosque and the Prophet's Mosque through Qeba Avenue. This has been achieved by fostering the place's vibrancy while emphasizing and promoting the cultural and Islamic values of the place.

Other projects are now in progress to make the city's roads more humane and transform them into vibrant pathways, such as Ohod Avenue, Al-Salam Avenue, Hijra Road, and the Historical Trail.

4.2 Ohod Avenue

Ohod Avenue is a path connecting the Prophet's Mosque, Sayyid al-Shuhada Square, and Mount Ohod, spanning almost 2.7 Kilometers in length, see Map 2. The project of improving Ohod Avenue is a part of the humanization program managed by the Al-Madinah Municipality. It aims to transform the avenue into an attractive hub for investments, improve infrastructure, enhance building facades, promote urban identity, invest in tourist attractions, and create a cohesive interface that showcases prominent landmarks for visitors. The project has focused on creating safe pathways for pedestrians, including a dedicated path for people with disabilities, as well as designated lanes for cars, buses, and bicycles. Additionally, it includes over fifteen thousand trees, specifically designed seating spots, and retail points. The implementation of this project is still ongoing. However, the Municipality has finished 80% of the project work [47].

Ohod and Qeba Avenues have common features that contribute to promoting livability and enhancing the quality of life in Al-Madinah city. However, Qeba Avenue has been completely transformed into an avenue exclusively for pedestrians, while Ohod Avenue still allows for motorized traffic. This has led to the need for dedicated lanes that run alongside the pedestrian and bike pathways. This exemplifies the broad variety of alternatives present at various levels to decrease reliance on cars and create an environment where walkers can move around freely.

4.3 Al-Salam Avenue

The Medina Region Development Authority [45] has improved Al-Salam Avenue as it is one of the major streets that connects the core district surrounding the Prophet's Mosque and other city districts. The main objective is to enhance the overall quality of life in the area, enhance the visual appeal, and make it easier for pilgrims and Umrah guests to move around and reach their destinations. The project encompasses pedestrian pathways, special paths for people with disabilities, watering points, and rest areas. Similar to Ohod Avenue, it features car lanes and parking facilities specifically for public transportation buses, see Fig. 3.

The implementation of this project is still ongoing. However, frequent visits, combined with the available visual resources, show the transformation of the street as it evolves into a vibrant place that prioritizes pedestrians and accommodates human needs.

4.4 Hijra Road

The Hijra Road Project spans a distance of 24 km within the city and traverses various tourist and historical attractions, such as Qeba Mosque, Qeba Avenue, Al Qaswa Park, historical farms, and residential districts. It encompasses the addition of pedestrian and cycling paths, as well as a specifically designed running track. The implementation of this project is still ongoing as shown in Fig. 4.

Fig. 3 Prioritizing pedestrians on Al-Salam avenue. *Source* [45]

Fig. 4 The ongoing transformation of Hijra Road towards a vibrant humanized avenue. *Source* [48]

4.5 Historical Trail

The historical trail is a futuristic initiative, spanning a 175 km route that will include over 40 historical and Islamic sites scattered around Al-Madinah. The path will be created to make it easier for visitors to explore these significant attractions, many of which are from the early days of Islam. This includes mosques where unique events occurred, such as the Qiblatain Mosque and Qeba mosque. Additionally, notable features encompass Mount Ohod, the Khandaq region, Othman's Well, and numerous prophetic monuments. The project will involve revitalizing these sites, as well as establishing public spaces, pedestrian pathways, and cultural activities to enhance the visitors' experience [45].

Overall, humanizing Al-Madinah's streets served as a fundamental component in improving and optimizing the economic and cultural growth of the region. This has been achieved by providing visitors with enriching and high-quality experiences in various areas by highlighting the city's relative advantage, cultural significance, and historical events, fulfilling visitors' interest in acquiring knowledge about it.

In addition, this humanization initiative in Al-Madinah main streets adopted several strategies, see Fig. 5, to create a vibrant, healthy, and resilient city including:

- **Prioritizing pedestrians:** Creating safe and attractive pedestrian-friendly infrastructure, with wider sidewalks, improved lighting, and dedicated walking zones to encourage active commutes and create a more vibrant public realm and healthy lifestyles.

Fig. 5 Humanizing strategies implemented in Al-Madinah's streets. *Source* The Authors

- **Providing sustainable mobility:** Adopting cleaner, more efficient public transportation systems and integrating bicycles, e-scooters, and other micro-mobility solutions to create a network of accessible mobility systems.
- **Enhancing urban experience:** Creating vibrant public spaces such as plazas, parks, and promenades along the streets to provide social gathering places for residents and visitors, and promote social interaction.
- **Improving visual identity**: Upgrading building facades and incorporating landscaping such as benches, bike racks, trash bins, public art installations, wayfinding signs, and lighting fixtures to enhance the functionality and attractiveness of streets for pedestrians with consideration to the local culture and traditions.
- **Increasing vegetation:** Introducing more green spaces, trees, plants, and landscaping elements along streets to improve air quality, reduce heat island effects, create visual appeal, and contribute to a healthier urban environment.
- **Promoting accessibility:** Building dedicated lanes for people with disabilities.

These strategies help humanize the city's streets and foster a more vibrant, healthier, and sustainable city, potentially reducing environmental and health impacts and promoting sustainable practices. This humanitarian effort serves as a model for other cities struggling with car-centric planning.

5 Discussion

It is crucial to discuss the humanization initiatives to ensure their effectiveness in response to the theoretical concepts of humanization and in achieving the Saudi Vision and goals. The documentary review shows that Al-Madinah Municipality has committed to developing and implementing humanization projects, considering the city's culture, religious, and historical significance. This, in turn, emphasized the city's qualities and values and improved the overall quality of life in several aspects including the health, the place's vibrancy, and the climate resiliency.

5.1 Al-Madinah as a Healthy City

Through the implementation of humanization efforts, Al-Madinah City is improving the physical and public health of its residents while also providing a more humane and responsive urban setting. Humanizing Al-Madinah's streets can contribute to public health in several ways:

- Enhancing the walkability and bikeability of a city promotes physical activity, leading to better public health.
- Reducing air pollution due to the decrease in the number of cars, resulting in cleaner air, and enhancing respiratory well-being.

- Enhancing air quality due to the increase in green areas and vegetation in open spaces.

It is worth mentioning that the World Health Organization (WHO) has officially recognized Al-Madinah City as one of the healthiest cities in the world, meeting global standards. The criteria included the achievement of objectives established by the Madinah Region Strategy Project and the implementation of a "Humanizing Cities" program [49].

5.2 *Al-Madinah as a Culturally Vibrant City*

The humanization initiatives successfully restored Al-Madinah's vibrancy, transforming it into an appealing tourist environment. Additionally, it prepared the city avenues to host a diverse range of social, cultural, creative, recreational, and tourist activities. This is due to a combination of culture, arts, and the human-environmental civilizational perspective, resulting in an enhancement of the overall quality of life. Humanizing Al-Madinah's streets can contribute to the city's vibrant culture in several ways:

- Preserving Al-Madinah city's religious identity and heritage as a result of the preservation of historical sites, incorporating the city's rich cultural heritage into urban design, and fostering cultural festivals.
- Promoting social interaction and the cultivation of a collective identity, thus fostering a feeling of responsibility toward the surroundings due to the vibrant streets with coffee shops, designated seating zones, and communal places.
- Providing a unique urban experience for millions of Muslims who are visiting Al-Madinah during the Umrah and Hajj seasons as a result of the expansion of public spaces and plazas.
- Promoting religious tourism through offering diverse cultural experiences for tourists in a pedestrian-friendly setting that links the significant Islamic landmarks in Al-Madinah City.

As a result, the humanization initiatives have accomplished remarkable successes both locally and globally. In 2018, two projects were awarded prestigious honors for their exceptional humanitarian and cultural contributions by the International Award for Livable Communities [50], recognized by the United Nations Environment Programme. Qeba Avenue project was one of the projects that received a silver award. Furthermore, in the same year, Qeba Street received the Tourism Excellence Award for Best Tourist Attraction Site, as well as the Tourism Excellence Award for the Best Cultural Festival in Saudi Arabia. Also in Kuwait, it was awarded the Best Municipal Initiative Award in the field of city humanization by the Gulf Cooperation Council [44]. The significance of these awards stems from the development approach to enhancing visual identity, as well as the preservation of culture and heritage through the implementation of sustainable practices. Furthermore, the awards also considered the discussion of the local community's involvement in development projects and the measurement of the overall quality of life in cities.

5.3 Al-Madinah as a Climate-Resilient City

A climate-resilient city is one that proactively adapts to the impacts of climate change. Humanizing Al-Madinah's streets can contribute to climate resiliency in several ways:

- Regulating temperature and mitigating the heat island effect as a result of increasing green spaces, street trees, and vegetation, while providing residents with natural areas for recreation.
- Reducing greenhouse gas emissions and cutting down carbon footprint as a result of improving the pedestrians' and biking infrastructure along with public transportation.
- Reducing stormwater runoff and flooding risks due to the replacement of impermeable surfaces such as "Asphalt" with green areas and street trees that allow rainwater to infiltrate the ground.
- Reducing the number of mobility trips and using spacious and pedestrian-friendly streets, near the holy sites (Qeba, Ohod, Al-Salam streets), to accommodate the anticipated increase in Hajj and Umrah pilgrims.

It is worth mentioning that Al-Madinah received recognition from the Sustainable Development Goals (SDG) Cities Program accredited by the UN-Habitat for its efforts as the first Saudi city, the first Arabic city, and the third city worldwide. Al-Madinah's progress from achieving the Silver Certificate in 2023 to achieving the Gold Level in 2024 demonstrates its prominent position at international and national levels in completing a comprehensive baseline data assessment of the city's progress in sustainable urban development with inclusive and transformative actions [45]. This is in line with Madinah's vision to create a city that focuses on the needs and well-being of its residents and visitors, with the goal of creating a vibrant community and enhancing the city's climate resiliency.

5.4 Al-Madinah Humanization Challenges

The humanization initiative has faced public criticism and opposition. While many visitors and residents appreciate the enhanced visual aesthetics, green areas, and streetscape elements, there is a significant number of locals who oppose the reduction of street width and the incorporation of wide sidewalks, as this restricts the available space for car traffic. Residents who still rely on private cars believe that promoting pedestrian movement comes at the cost of noticeable impacts on car movement, trip length, congestion levels, and overall difficulty of travel by cars, particularly during peak hours. Therefore, it is necessary to involve the local community of Al-Madinah in the planning process through participatory design workshops, feedback sessions, surveys, and public consultations to ensure that the humanization efforts align with the needs and preferences of all residents. Educating the public and prompting travel behavior change is a significant factor that leads to public acceptance.

Furthermore, most efforts primarily focus on particular streets, specifically surrounding the Prophet's Mosque, while numerous regions and roads inside the city remain undeveloped or have yet to undergo development. Hence, residents need to witness ongoing enhancements in various areas and remain informed about them.

Ultimately, this project has effectively maintained the overall cultural essence, although there is a common concern in Saudi Arabia about unintentionally imitating the European and Western approach to humanization without adequately incorporating the local culture. The Saudi endeavors show significant opportunities for innovation under the scope of humanization within the local culture theme.

6 Conclusions

In conclusion, Al-Madinah, an originally car-centric city, has faced challenges like traffic congestion, air pollution, and limited pedestrian access. Inspired by Saudi Vision 2030's focus on improving quality of life, the city is undergoing a significant transformation of humanizing its streets. Humanizing Al-Madinah's streets has revitalized the city by prioritizing pedestrians, creating a sustainable transportation network, enhancing public spaces, and incorporating vegetation and landscaping. These efforts not only improved public health, cultural life, and the city's climate resilience, and thus the overall quality of life, but also promoted religious tourism by offering diverse cultural experiences for tourists in a pedestrian-friendly setting that links the significant Islamic landmarks in Al-Madinah City

However, challenges still need to be addressed, including society's resistance to reduced car traffic and the need for wider development across the city. Through these efforts, Al-Madinah has demonstrated a commitment to the broader goals of the Saudi Vision 2030. By emphasizing humanization, Al-Madinah can serve as a model for other cities to change from long-term car-centric planning to vibrant, healthy, and resilient urban development.

The implications of this transformation of Al-Madinah include reducing Carbon footprint, mitigating the urban heat-island effect through reducing car dependency, the increased incorporation of green spaces, trees, and better-designed urban landscapes. Additionally, this transformation has significant positive implications for public health including increasing residents' physical activities, improving well-being, and enhancing social cohesion through prioritizing pedestrian-friendly infrastructure, such as wider sidewalks and dedicated pedestrian zones, integrating green spaces and public areas into urban design, and creating human-centric communal spaces.

References

1. Dahim MA (2023) Challenges and opportunities in development sustainable transportation system in Saudi Arabia. Appl Eng Technol 2(1):34–45
2. Aldegheishem A (2024) Assessing urban sustainability in Saudi Arabia: an empirical evidence from Al-MedinaAl-Munawwarah. Environ Res Commun 2024:1
3. Badawi S, Farag AA (2021) Young Saudi women's travel behavior change over 2015/2020. J Transp Health 21:101080
4. Ministry of Municipal and Rural Affairs (2019) Future Saudi cities programme-city profiles series: Madina. Available at https://unhabitat.org/sites/default/files/2020/04/madinah.pdf
5. Ministry of Municipal and Rural Affairs (2019) Urban prosperity report (Al-Madinah Al-Munawarh). Available at: https://saudiarabia.un.org/sites/default/files/2020-12/%D8%AA%D9%82%D8%B1%D9%8A%D8%B1%20%D8%AD%D8%A7%D9%84%D8%A9%20%D8%A7%D8%B2%D8%AF%D9%87%D8%A7%D8%B1%20%D8%A7%D9%84%D9%85%D8%AF%D9%86%20-%D8%A7%D9%84%D9%85%D8%AF%D9%8A%D9%86%D8%A9%202019%20%D8%A7%D9%84%D9%85%D9%86%D9%88%D8%B1%D8%A9.pdf
6. Doheim RM, Farag AA, Badawi S (2019) Smart city vision and practices across the Kingdom of Saudi Arabia—a review. Smart Cities Issues Challenges 2019:309–332
7. Faheem HB, Shorbagy AME, Gabr ME (2024) Impact of traffic congestion on transportation system: challenges and remediations—a review. Mansoura Eng J 49(2):18
8. Malqarni SS (2016) A review of prevalence of obesity in Saudi Arabia. J Obes Eat Disord 2:2. https://doi.org/10.21767/2471-8203.100025
9. World Health Organization (2002) Physical inactivity a leading cause of disease and disability. Online. https://www.who.int/news/item/04-04-2002-physicalinactivity-a-leading-cause-of-disease-and-disability-warns-who
10. Yang M, Bing Z, Perlman J, Pratt AC (2016) Humanizing cities through culture. UNESCO. Available at: https://unesdoc.unesco.org/ark:/48223/pf0000260634
11. Doheim RM, Farag AA, Badawi S (2020) Success measures for transforming into car-free cities: recommendations for implementation. In: Humanizing cities through car-free city development and transformation, pp 231–267
12. Gehl J (2010) Cities for people. Island Press, New York
13. Lynch K (1981) Good city form. MIT Press, New York
14. Jacobs J (1961) The death and life of great American cities. Vintage Books, New York
15. Badawi S, Farag AA (2023) Improving the urban qualities of historic streets in Jeddah. Hist Environ Policy Pract 14(1):30–57
16. Jeekel H, Emanuel M, Oldenziel R, Schipper F (2015) What is happening to cars in urban spaces? Database 313(166):53
17. Lacey GC (1975) The dominant car. In: Metropolitan transport 1975: The way ahead? Preprints of papers: The way ahead? Preprints of papers. Institution of Engineers, Sydney, pp 12–16
18. Hynes M, Seoighthe E (2018) Heading in the right direction? Investigating walkability in Galway City, Ireland. Urban Sci 2(2):31. https://doi.org/10.3390/urbansci2020031
19. Hynes M (2017) At a crossroads: investigating automobility and its implications for local urban transport policy design. Urban Sci 1(2):14. https://doi.org/10.3390/urbansci1020014
20. Szell M (2018) Crowdsourced quantification and visualization of urban mobility space inequality. Urban Plan 3(1):1–20. https://doi.org/10.17645/up.v3i1.1209
21. General Authority of Statistics (2017) Housing survey. Available online at: https://www.stats.gov.sa/ar/node/14481. Accessed date: 23.1.2021
22. Guerrero-Ibáñez A, Flores-Cortés C, Damián-Reyes P, Andrade-Aréchiga M, Pulido JRG (2012) Emerging technologies for urban traffic management. Urban Develop 59:1
23. Farag AA (2022) How sustainable is mobility in cities branded the happiest? Linking sustainability and happiness: theoretical and applied perspectives, pp 151–174
24. Douglas MJ, Watkins SJ, Gorman DR, Higgins M (2011) Are cars the new tobacco? J Public Health 33(2):160–169

25. Alsawas AT, Sayed KMS (2021) Theories of new urbanism and the future of urban development in KSA (a case study of Al-Madinah Al-Munawwarah city). Umm Al-Qura Univ J Eng Archit 12(1):1
26. Tayan O, Alginahi YM, Kabir MN, Al Binali AM (2017) Analysis of a transportation system with correlated network intersections: a case study for a central urban city with high seasonal fluctuation trends. IEEE Access 5:7619–7635
27. Aljoufie M (2021) Toward an effective analysis of public transportation demand factors in car-dependent cities: case of Makkah City, Saudi Arabia. Urban Plan Transp Res 9(1):257–275
28. MMM Group, Moriyama & Teshima Architects and Planners (2011) Environment Plan, Comprehensive Plan for Madinah
29. Bazzaz AA, Al-Juhani AK, Al-Erwi NS, Kurdi KN (2011) The air pollution concern within the holy prophet mosque in Al-Madianh Al-Munawarah. J Environ Sci Eng 5(5):1
30. Al-Jeelani HA (2012) Trends of ambient concentrations of gaseous air pollutants in Almadinah Al Menawwarah, Central Area, Saudi Arabia. JKAU Met Environ Arid Land Agric Sci 23(2):31–53
31. Schneider F, Ortiz J, Vanos J, Sailor D, Middel A (2022) Evidence-based guidance on reflective pavement for urban heat mitigation: a case study in phoenix. Arizona. https://doi.org/10.21203/rs.3.rs-1965978/v1
32. Farag AA, Doheim RM, Badawi S (2019) Evaluating heat island effect at university campus with reference to LEED v4. Resourceedings 2(1):01–13
33. Abdou AEA, Turki HM (2014) Variation of surface temperature over AL-Madinah Al-Munawarah, Saudi Arabia, (1978–2013). Int J Sci Res 3:1734–1751
34. Alahmadi RN (2014) Modelling the relationship between pilgrims' pedestrian casualties and land use type: a case study of Al Madinah Al Monawarah. Doctoral dissertation
35. Frank L, Mayaud J, Hong A, Fisher P, Kershaw S (2019) Unmet demand for walkable transit-oriented neighborhoods in a midsized Canadian community: market and planning implications. J Plan Educ Res 42(4):568–584. https://doi.org/10.1177/0739456x19831064
36. Liao B, Berg P, Wesemael P, Arentze T (2020) How does walkability change behavior? A comparison between different age groups in the netherlands. Int J Environ Res Public Health 17(2):540. https://doi.org/10.3390/ijerph17020540
37. El Ela ABU, Samy H (2016) Monitoring some smart city geographical characteristics of medina in Saudi Arabia. Roman J Geography R Roumaine Géographie 60(2):1
38. Jung E, Kim Y (2018) Commuting, communication, and community engagement from a communication infrastructure theory perspective. Commun Res 48(5):740–766. https://doi.org/10.1177/0093650218801745
39. Wang Y, Sousa-Poza A (2018) Commute time and subjective well-being in urban China. China Econ Rev 48:188–204. https://doi.org/10.1016/j.chieco.2016.03.002
40. Abdi M, Soltani A (2022) Which fabric/scale is better for transit-oriented urban design: case studies in a developing country. Sustainability 14(12):7338. https://doi.org/10.3390/su14127338
41. Doheim RM, Farag AA, Kamel E (eds) (2020) Humanizing cities through car-free city development and transformation. IGI Global, New York
42. MOMRAH (2024) Humanizing cities. Ministry of Municipal and Rural Affairs and Housing. Available at: https://momrah.gov.sa/en/initiatives/2658
43. QoL Program (2023) Quality of life program-implementation plan (2021–2025). Kingdom of Saudi Arabia 2030 vision. Available at: https://www.vision2030.gov.sa/media/rlrhwdlg/2021-2025-quality-of-life-program-delivery-plan-en.pdf
44. Mughairy et al (2024) Cities for people towards vibrant public places. The Arab urban development institute (3rd). Available at: https://araburban.org/wp-content/uploads/2024/03/issue-3-2-en.pdf
45. Al-Madinah Region Development Authority (2024) Al Madinah humanization program. Available at https://www.mda.gov.sa/projects. Accessed in May 15, 2024
46. Suliman H (2021) The role of sense of place in place-making and urban conservation for heritage sites: Al-Madinah Al-Munawara case study. Eng Res J 172:85–120. https://doi.org/10.21608/ERJ.2021.208234

47. Amana.al-Madinah (2024) City municipality projects platform. Available at https://geoservices.amana-md.gov.sa/madina-projects/project-details/1081
48. Med_Projects (2024) Available at: https://www.instagram.com/p/Cz3_XYxIFRD/?img_index=4
49. Desk W (2021) WHO includes Madina among world's healthiest cities. GeoNews. Available at https://www.geo.tv/latest/331661-who-includes-city-of-madinah-as
50. International Award for Livable Communities (2018) Winners 2018. Available at: http://www.livcomawards.org/2018-awards/winners.html

The Potential of Biophilic Urban Planning for Climate Resilience Cities in Africa: The Case of Dakar Metropolitan Region

Cheikh Cisse

Abstract In the face of the challenges posed by climate change and rapid urbanization, particularly in the Global South, it is essential to create cities resilient to the complex impacts of climate change. This chapter examines the case of Dakar and other African examples to explore the integration of biophilic design principles into urban planning and construction practices as a novel approach to fostering healthy buildings and communities within climate-resilient cities. Drawing on studies and concrete experiences, the chapter demonstrates that biophilic design, inspired by the intrinsic connection between humans and nature, offers a promising pathway toward sustainable urban development. This study presents practical examples and real-world projects in the Dakar region, as well as other African contexts, illustrating a paradigm shift toward more sustainable and climate-resilient cities. It also highlights innovative design and planning strategies that incorporate biophilic elements such as green roofs, vertical gardens, and natural lighting to improve indoor air quality, reduce energy consumption, and promote well-being. Finally, the chapter discusses the potential of biophilic design to create healthier and more resilient urban environments, with a particular focus on Dakar, and addresses its impact on community well-being, energy efficiency, and ecological sustainability, while explaining how these strategies can be extended and adapted to different urban contexts, particularly in Africa.

Keywords Biophilic design · Resilient cities · Climate change · Healthy buildings · Urban planning · Sustainable architecture

---

C. Cisse (✉)
African Studies Centre Leiden, Leiden University, Leiden, The Netherlands
e-mail: cisse27@yahoo.fr

Ecole Supérieure des Professions Immobilières, Paris, France

A. Cheshmehzangi et al. (eds.), *Designing Healthy Cities*, Urban Sustainability,
https://doi.org/10.1007/978-981-96-5538-0_4

1 Introduction

On a world scale, rapid urbanization, combined with exponential population growth concentrated in urban areas, is creating major climate challenges to be met. According to United Nations estimates, by 2050, almost 70% of the world's population will be living in urban areas, with a large proportion in major African cities [27]. This urbanization puts great pressure on cities' infrastructure, natural resources, and governance systems, increasing their vulnerability to climate impacts.

Africa's major cities, such as Dakar on a metropolitan scale, illustrate this high level of urbanization and the urgent challenges posed by climate change. According to the African Development Bank's [2] State of the Economy in Africa report, published in 2022, the African continent, with its sustained and uncontrolled urbanization, is particularly vulnerable to the effects of climate change, with potentially destructive consequences for its populations and rapidly expanding cities, such as more frequent droughts and floods.

Faced with these challenges, urban resilience is an imperative for political decision-makers and urban professionals alike. Cities are called upon to adapt to the effects of climate change while ensuring economic growth, public health, and quality of life for their inhabitants. It is in this context that biophilic urban design is positioned as a promising solution among others.

Inspired by the forms, patterns, and processes of nature, biophilic design offers an innovative approach to designing climate-resilient urban spaces. By integrating natural elements such as vegetation, water, and light into urban planning and building design, biophilic design, as demonstrated in the main body of the chapter, can improve air quality, reduce urban heat islands, and promote the well-being of residents. It's an approach that can reduce urban temperatures, lower stress levels, and improve the mental health of city dwellers.

1.1 Study Area: Dakar on a Metropolitan Scale

Dakar and its hinterland, representing the Dakar metropolitan region, face major climatic challenges exacerbated by rapid and poorly planned urbanization. Rising temperatures, coastal erosion, and extreme weather events such as recurrent flooding make the region particularly vulnerable. These climatic challenges add to existing pressures on urban infrastructure, housing, health services, and poor urban waste management.

The suburbs of Dakar, including towns such as Pikine, Guédiawaye and Rufisque, regularly experience severe flooding during the rainy season, causing major damage to homes and infrastructure, and endangering the health of residents. Erosion, particularly visible in the coastal areas of Rufisque, Yoff, and Bargny, threatens habitats and local economic activities, notably artisanal fishing (Map 1).

Map 1 Dakar metropolitan area. *Source* Open Street Map, Cheikh CISSE rehabilitation (2024)

Faced with these many challenges, mainly climatic, it is important to ask how biophilic urban design could offer innovative solutions to mitigate these impacts by integrating green spaces, ecological corridors and green infrastructure into metropolitan Dakar.

1.2 Plan

In this chapter, we explore the potential of biophilic design to address climate and urban challenges in Africa, highlighting specific examples of projects and initiatives at the scale of metropolitan Dakar. By examining how biophilic design can contribute to urban resilience, the chapter provides insights to guide the future development of Dakar and, beyond, Senegalese and African cities towards a sustainable, climate-resilient future.

2 State of the Art

2.1 *Foundations of Biophilic Design*

Biophilic design represents an innovative approach to urban design, based on fundamental principles aimed at harmoniously integrating natural elements, in connection with humans, into built environments. These foundational principles are essential for understanding and effectively applying biophilic design in urban contexts, particularly in African cities such as Dakar, which is significantly affected by the impacts of climate change.

2.1.1 Definition and Fundamental Principles

Biophilic design, as defined by Kellert [14] as "the inherent human inclination to affiliate with nature and life processes," is based on several fundamental principles. These include the connection with nature (flower beds, bird feeders, butterfly gardens, potted plants), the presence of organic forms, the use of natural elements (daylight), vegetation (green walls or roofs), water (ponds, aquariums), and the promotion of biodiversity. Similarly, Heerwagen [11] suggests that biophilic design aims to "maintain, enhance, and restore the beneficial connections between humans and nature in the built environment." The goal is to create spaces that reflect natural processes and systems, thereby promoting human well-being. According to a study by Terrapin Bright Green [24], biophilic design should offer a multisensory experience by stimulating not only sight but also hearing, smell, touch, and even taste, in order to reconnect individuals with nature in a holistic manner.

As an example, vertical gardens on urban building facades offer numerous advantages. They increase biodiversity by creating habitats for flora and fauna, improve air quality by absorbing CO_2 and filtering fine particles, and reduce urban heat islands through shading and the evapotranspiration of plants [3]. The use of natural materials like wood also has beneficial effects. Trees absorb CO_2 and release oxygen as they grow through photosynthesis, storing carbon. Once transformed into wood, this carbon remains trapped in building structures for the long term. Replacing every felled tree with a new one maintains this virtuous cycle of CO_2 capture, contributing to a sustainable carbon sink when the wood is sourced from responsibly managed forests [9]. The presence of water in indoor spaces, such as fountains or water walls, has a calming effect due to the natural sounds it produces. The [24] study notes that fountains create a microclimate that cools the space with their sounds. Tiles and railings near waterways cool the feet and hands thanks to the conductivity of the materials used.

2.1.2 The Human-Nature Connection: Inspiration for Biophilic Design

The intimate relationship between humans and nature forms the central foundation of the concept of biophilia. Researchers like [29] have highlighted the vital importance of this innate connection with nature for human well-being, introducing the biophilia hypothesis, which posits that humans possess a natural affinity for nature. Kaplan [13] developed the Attention Restoration Theory, demonstrating that exposure to nature restores cognitive abilities. By incorporating natural elements into urban environments, biophilic design recreates this connection and offers numerous benefits in urban design.

Integrating natural elements into urban environments not only re-establishes this essential connection but also brings significant psychological, social, and health benefits. In the workplace, for instance, the presence of vegetation in office spaces, especially in urban settings, has been correlated with a notable reduction in stress and anxiety, with a 20% decrease in cortisol levels observed in employees exposed to plants. This presence of vegetation also enhances concentration, engagement, and employee productivity, which can increase by up to 8%. Additionally, improvements of 13% in well-being, creativity, and cognitive abilities have been reported, while reducing absenteeism and attracting new talent [16].

The benefits of biophilic design also extend to public health. Exposure to natural elements in urban areas has been associated with lower rates of depression, making biophilic design an essential tool for urban planners aiming to improve public health. Moreover, urban green spaces play a key role in improving air quality. Trees and plants absorb pollutants such as nitrogen dioxide and fine particles, often responsible for respiratory problems like asthma and bronchitis. Thus, in addition to strengthening the human-nature connection, integrating biophilic elements into urban design significantly contributes to promoting the physical and mental well-being of urban residents by fostering a healthier and more balanced environment.

2.2 *Integrating Biophilic Design into Urban Planning*

Biophilic urban planning involves integrating the principles of biophilic design into the urban development process. This approach can play an important role in creating resilient and sustainable neighborhoods in the face of climate change challenges caused by rapid urbanization, particularly on the African continent.

2.2.1 Biophilic Urban Design Strategies

Biophilic urban planning strategies, such as the integration of parks and green spaces, the creation of ecological corridors, the use of natural materials in construction, and the implementation of ecological stormwater management systems, aim to promote biodiversity, mitigate urban heat islands and improve residents' quality of life. The integration of parks, along the lines of Central Park in New York, Les Buttes Chaumont in Paris or Parc de Hann in Dakar, provides spaces for relaxation and recreation while promoting biodiversity and reducing stress and anxiety among residents [4]. The creation of ecological corridors enables the free movement of species and connects green spaces, thus contributing to urban biodiversity [23].

The use of natural materials in construction, such as certified wood and local materials, as illustrated by the Bullitt Center in Seattle, reduces the carbon footprint of buildings and creates an aesthetic connection with nature [17]. Finally, green stormwater management systems, such as rain gardens and green roofs, help capture and filter runoff, reducing flood risk and improving urban resilience [5].

These different strategies demonstrate the potential of biophilic design to transform urban environments into more sustainable and resilient spaces.

2.2.2 Potential Impact of Biophilic Design on Urban Quality of Life and Resilience to Climate Change

Integrating biophilic design into urban planning can have a significant impact on the quality of life of city dwellers and the resilience of cities to climate change. By providing accessible green spaces, such as community parks and gardens, biophilic design improves the mental and physical health of residents, reducing stress and encouraging outdoor activities [13, 26]. Ecological corridors, which link different natural spaces, help maintain urban biodiversity and facilitate the movement of wildlife, thus contributing to the preservation of ecosystems [4].

In addition, green infrastructure such as vegetated roofs and walls, rain gardens and vegetated retention basins play a major role in urban thermal regulation, reducing urban heat islands and improving thermal comfort [6]. These infrastructures also contribute to sustainable stormwater management, reducing the risk of flooding and improving water quality [10]. By providing natural protection against natural disasters such as flooding, biophilic design enhances the resilience of cities to the effects of climate change [15]. Thus, integrating biophilic design into urban planning is essential to developing more resilient, sustainable, and liveable cities. It contributes to improving the quality of life of residents and reinforces the ability of cities to adapt and withstand climate challenges [14] (Newman & Beatley 2006).

3 Methodology

For the development of this chapter, a rigorous and comprehensive research methodology was employed, integrating both a systematic literature review and empirical methods to collect qualitative and quantitative data.

Firstly, a systematic review of the literature was conducted using recognized academic databases such as Google Scholar, PubMed, ScienceDirect, and ResearchGate. The search terms used included "urban biophilic design," "urban biodiversity in

Dakar," "urban forests in Africa", and "urban ecological corridors". To ensure a representative and relevant sample of studies, the criteria for selecting articles focused on their relevance to the study objectives, their publication date to ensure the currency of the information, the quality of the publications (favoring studies published in peer-reviewed scientific journals), and access to the full text for in-depth analysis.

The selected articles underwent a systematic content analysis to identify recurring themes and emerging trends in the literature. This analysis revealed common patterns such as the impact of green spaces on residents' well-being, the ecological benefits of green roofs and vertical gardens, as well as the challenges associated with integrating nature into African urban environments. A comparative analysis was then used to evaluate different approaches to biophilic design in various urban contexts across Africa, highlighting the successes and limitations of existing initiatives. Finally, a narrative synthesis was performed to integrate the qualitative and quantitative data, providing a coherent and comprehensive interpretation of the results.

In addition to the literature review, semi-structured interviews were conducted with seven urban planning and environmental experts in Dakar. These experts were selected based on their experience and expertise in implementing urban projects. The interviews, which lasted an average of thirty-five minutes, were conducted in person or by phone, depending on the participants' availability. The questions addressed topics such as the application of biophilic principles in urban projects, the impact on residents' quality of life, and the challenges encountered when integrating nature into urban environments.

Alongside, the study drew upon our direct observation and lived experience, having grown up in Dakar and being well-acquainted with the city. This approach allowed us to leverage an in-depth knowledge of local urban dynamics, environmental practices, and residents' expectations regarding green spaces and urban sustainability. The observation was systematically conducted across different neighborhoods of Dakar, taking into account the geographical, socio-economic, and cultural diversity of the metropolis. This enabled the capture of nuances and realities often overlooked in academic studies, thereby enriching the analysis with concrete empirical perspectives.

The data collected from the semi-structured interviews and direct observations were analyzed using a systematic approach to ensure the rigor and relevance of the conclusions. Each interview was transcribed and thoroughly analyzed to identify recurring themes, differences in opinion, and shared experiences among the consulted experts. A thematic analysis was carried out to identify the main emerging patterns regarding the application of biophilic principles in urban projects in Dakar, particularly the impacts on residents' quality of life and the challenges faced in integrating nature into urban environments.

Direct observation and lived experience helped contextualize these results, offering a perspective rooted in the reality of Dakar's urban dynamics. This empirical approach enriched the thematic analysis by adding nuanced details about local environmental practices and residents' expectations regarding urban sustainability. The observations were systematically noted and compared with interview data to verify the consistency of the information collected.

The results were then interpreted to identify key trends and highlight the specific challenges in Dakar regarding the integration of biophilic design into urban planning. This analysis led to the formulation of recommendations tailored to the local context, considering the socio-economic and cultural realities specific to Dakar. This methodological approach, which combines a rigorous documentary analysis with empirical methods based on observation and personal experience, formed the foundation for this chapter, providing a solid basis for recommendations on biophilic urban planning in African contexts.

4 Results: A Few Examples of Biophilic Attributes in the Senegalese and African Urban Contexts

4.1 Urban Forests: Biodiversity on the Dakar Regional Scale

The Dakar region is rich in major urban forests, offering exceptional potential for the development of biophilic design in urban planning. The use of these urban forests as central elements of urban planning not only preserves biodiversity and reduces urban heat islands, but also promotes the physical and mental well-being of residents by providing spaces for relaxation and contact with nature. Each of these major forest areas has its own particularities:

The Park of Hann lies at the heart of the forest ecosystem that ensures the well-being, shelter, and development of plants, animals, and people [28] in the Dakar region. Covering an area of 60 hectares, Hann Park is located 6 km from the downtown of Dakar. It represents one of Dakar's largest urban forests and contributes to Senegal's growing biodiversity heritage. Against a backdrop of urbanization and strong demographic growth, Hann Park helps to maintain the region's socio-ecological balance [12]. Its green spaces and water features filter atmospheric pollutants, as one hectare of urban forest can absorb up to 15 tonnes of fine particles per year (Constructive [7]), while producing oxygen and reducing urban heat islands through evapotranspiration and shading. However, since 2005, Hann Park has been subject to flooding, threatening plant species not adapted to hydric conditions. Irregular construction in the vicinity, intended to act as a buffer zone, is the main cause of flooding. The degradation of plant cover and the effects of climate change are affecting biodiversity throughout the Sahel region, according to [20]. Consequently, there is a need to protect this ecological entity to maintain urban biodiversity and enable the park to fulfill its ecological functions [8].

The corniches of Dakar: from 1943 onwards, the colonial administration recognized the importance of using brushwood and plantations of various species to stabilize and beautify the maritime corniches of Dakar, located to the east and west of the Senegalese capital. This initiative led to the classification and intensive planting of the corniches, particularly to the east. In addition, a project to revitalize the western corniche incorporates 7 km of green promenades and recreational areas along the coastline, enabling residents to enjoy panoramic ocean views, while strolling and engaging in physical activities such as walking and jogging. This project illustrates the city of Dakar's commitment to creating biophilic public spaces [18]. However, the project suffers from a lack of regular maintenance. In fact, between the excessive construction on the so-called "maritime" domain, formally prohibited, and the accumulation of plastic waste and other detritus detracting from the attractiveness of these public spaces, these cornices "die" over time.

Mbao urban forest: classified in 1940 and initially covering 770 hectares, the Mbao forest was reduced to 500 hectares in 1972 to allow the creation of the Dakar industrial zone. Today, the forest is threatened by the growing needs of the local population due to heavy urbanization and a lack of management. It is also affected by growing insecurity. Preserving and securing the forest is urgent and essential if it is to continue to fulfil its ecological and social functions, and provide valuable goods and services to populations in search of green spaces.

Coastal areas: in 1950, over 4000 hectares of Dakar's sandy coastline (Malika, Retba, Cambérène) were reforested to stabilize dunes and protect coastal systems, particularly market-garden basins. However, over 2000 hectares of these restored areas have been urbanized (Cambérène and Yoff). Nevertheless, over 1000 hectares of filao plantations remain, constituting valuable peri-urban forest areas located less than 30 km from the city [21]. These plantations are under increasing land pressure and are often razed for new housing developments, notably in Guédiawaye and Malika, posing a serious threat by increasing the risk of coastal erosion.

The peripheral green belt, which could form part of Dakar's peri-urban forest system, includes the Deny-Youssouph forest (400 ha), the Sébikhotane forest (520 ha), and the Dialao hills. The latter are also threatened by the extraction of building materials, as well as growing pressure on land from earthworks and housing developments (Map 2).

Dakar's urban forests, while valuable for their biodiversity, ecological services, and potential for biophilic urban design, are threatened by urbanization and require sustainable protection and management to preserve their essential functions.

4.2 *Biophilic Design: A Method for Energy Efficiency and Reducing Greenhouse Gas Emissions (GES)*

Biophilic design can have a significant impact on buildings and communities in the Dakar region by promoting residents' well-being, enhancing energy efficiency, and strengthening community resilience to climate change. Indeed, it plays a crucial role in improving the energy efficiency of urban buildings and reducing greenhouse gas emissions. The use of natural light reduces electricity consumption related to artificial lighting, thanks to well-placed windows and high-performance glazing. Similarly, natural ventilation reduces the need for air conditioning in summer, through strategic openings and carefully designed ventilation systems. These two choices decrease

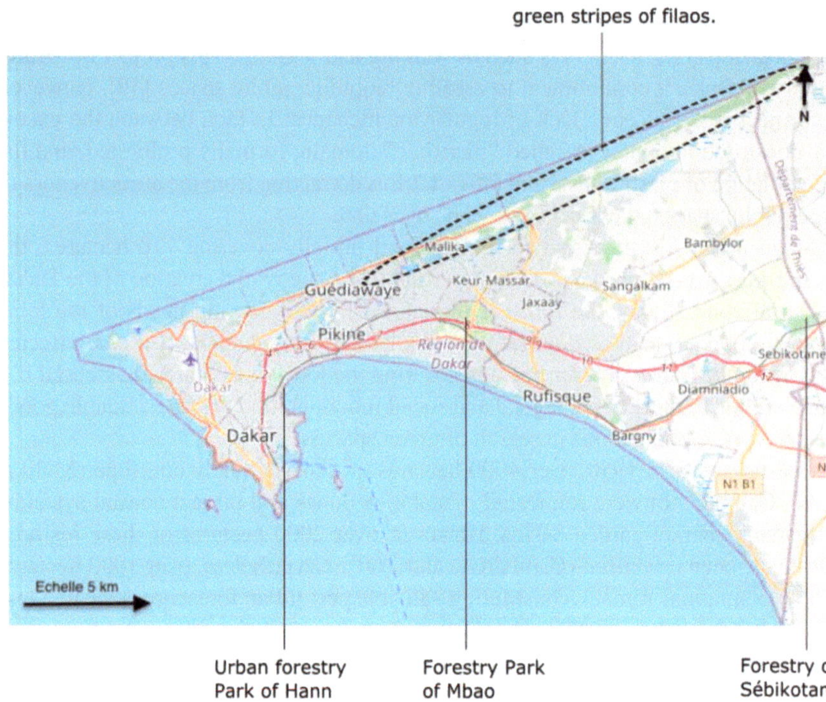

Map 2 Main forest areas in the Dakar region. *Source* Open Street Map, Cheikh CISSE rehabilitation (2024)

dependence on non-renewable energy sources and reduce the carbon footprint of buildings [25].

For instance, a study shows that greening buildings through the evapotranspiration of plants on roofs, terraces, and facades reduces the surrounding air temperature. This process uses energy to convert water absorbed by the plants into vapor, thereby lowering urban summer temperatures. More specifically, green facades help regulate building temperatures by reducing surface temperatures by 7–15 °C [1].

However, despite these potential benefits, our observations indicate that such solutions remain marginal in Dakar's urban landscape, where the majority of new constructions, particularly large-scale ones, do not sufficiently integrate biophilic design to enhance energy efficiency and reduce greenhouse gas emissions. This is due, according to the experts consulted, to a lack of technical expertise, regulatory incentives, and financial means to implement these forms of construction. Solutions such as the integration of planted courtyards, skylights, verandas, and shed roofs, combined with automated management systems, would optimize light intake while limiting urban overheating.

Beyond these environmental benefits, biophilic design plays an important role in public health. By promoting the presence of nature in urban spaces, it helps reduce levels of psychological disorders, thereby improving residents' quality of life. Public health and environmental experts agree that the temperature reduction due to greening could also have beneficial effects on physical health, notably by reducing the risks of cardiovascular and respiratory diseases, often exacerbated by excessive heat. Greened environments also help reduce incidents of heat-related illnesses, such as heatstroke, dengue fever, and dehydration, which are of particular concern in the context of climate change in Dakar.

These elements, combined with the positive effects on energy efficiency, make biophilic design an essential holistic approach for sustainable development in African urban contexts.

4.3 *Ecological Corridors in the African Urban Context*

The establishment of ecological corridors represents an innovative and essential response to the degradation of urban and rural ecosystems. These initiatives reflect a significant shift in the way societies approach environmental management, with an emphasis on restoring and enhancing natural spaces at the very heart of urbanized areas.

In Algiers, Algeria, the Oued El Harrach renaturation project, part of the Algiers 2029 Strategic Plan, embodies this vision by aiming to restore and revitalize a once-degraded watercourse [19]. This approach is part of a wider strategy to preserve urban biodiversity and create ecological corridors within the city. The aim is to restore the ecological functions of this river ecosystem, thereby promoting ecological connectivity between natural habitats and providing movement opportunities for urban flora and fauna.

Similarly, in South Africa, the Wild Coast illustrates how local arrangements are emerging to reconcile environmental conservation with the economic imperatives of coastal tourism. This patrimonialized ecological front bears witness to a collective desire to preserve natural resources while sustainably exploiting tourism activities [22].

In addition, the Great Green Wall, initiated by the African Union in 2007, is an ambitious project to create an 8000 km ecological corridor across more than 11 Sahelian countries, including cities such as Dakar, Niamey, N'Djamena and Khartoum. The project combats desertification, restores degraded ecosystems and improves local living conditions by enabling the connectivity of natural habitats for plant and animal species, sequestering carbon, managing land sustainably, and creating green jobs. In Dakar, as developed earlier in the text, this includes the planting of millions of trees and the development of urban forest parks, enhancing biodiversity, and providing green spaces for residents.

Despite promising initiatives, the implementation of ecological corridors in Africa is often hampered by limited resources, land constraints, demographic pressures, and economic priorities. Nevertheless, these projects, along with the potential of forest massifs and biophilic building design, demonstrate a growing awareness of the importance of these different approaches to biodiversity preservation and ecosystem resilience. They offer valuable models to inspire future initiatives on the continent, notably by integrating biophilic design into urban planning to mitigate the effects of climate change and improve residents' quality of life. By optimizing natural resources and promoting a healthier environment, these projects contribute to building more sustainable and resilient cities for future generations.

5 Discussion: Prospects and Recommendations

5.1 Potential for Expansion and Adaptation of Biophilic Design to Other Urban Contexts

The potential for expansion and adaptation of biophilic design to other urban contexts in Senegal and Africa is significant, offering opportunities to improve the quality of life of inhabitants and strengthen the resilience of cities in the face of climatic challenges.

Given the strong urbanization underway in Senegal, the Dakar-Thiès-Mbour triangle system, formed by these three cities located less than 80 km apart, is destined to become a mega-agglomeration in the next 50 years due to the advanced metropolization and sprawl of the Dakar region. This triangle includes 10,000 hectares of classified forests (Pout, Thiès, Bandia, Nianing, and Pointe Sarène), which is essential to preserve. Management measures, such as filling voids, restoring natural stands, and equipping forests, need to be taken now. The tourism importance of the Petite Côte and the considerable future needs require immediate action. The efforts of the Société d'Aménagement de la Petite Côte (SAPCO) in terms of mass planting, ornamental planting, and roadside planting illustrate the local capabilities and existing potential (Map 3).

In Nairobi, Kenya, the "Nairobi River Regeneration" project aims to restore urban river ecosystems, thereby helping to regulate rainwater and preserve local biodiversity. The project encompasses a number of key initiatives, such as the decontamination of river water, the reforestation of riverbanks with native plant species, and the creation of recreational areas along the banks. In addition to improving water quality, these efforts help prevent urban flooding by increasing water infiltration and reducing surface runoff. In parallel, the regeneration of the river enables the reintroduction of local flora and fauna, creating natural habitats in the midst of an urban environment. The green spaces created along the river provide residents with areas in which to relax, stroll, and encounter nature; thereby promoting their mental and physical well-being. These interventions benefit not only the environment but also the local community, promoting public health and strengthening the bond between residents and their natural environment.

These examples thus demonstrate the ability of biophilic design to adapt to local specificities while offering innovative solutions to urban challenges in Africa.

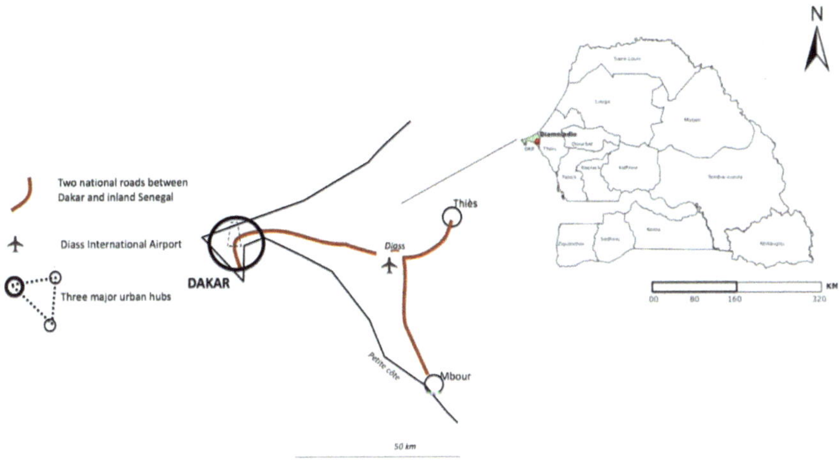

Map 3 The metropolitan future: three major conurbation urban poles. *Source* Realized by Cheikh CISSE (2024)

5.2 *Additional Research Needs for a Biophilic Urban Future*

In line with the analyses presented, further research and the development of robust public policies are crucial to fully realize the potential of biophilic urban design. Large-scale, long-term studies are needed to better quantify the impacts of biophilic design on urban residents' physical and mental health, biodiversity within cities, and overall resilience to climate change. Such research would consolidate the scientific evidence base, guiding best practices in biophilic design implementation.

For the Dakar metropolitan area, conducting macro-scale research into residents' perceptions and prioritizing biophilic urban design is particularly relevant given the diverse impacts of climate change. Surveys and field studies could provide invaluable insights into the expectations and needs of local communities, enabling the development of policies that are finely tuned to specific local contexts. To encourage the widespread adoption of biophilic design, it is essential to establish a strong regulatory framework and economic incentives. Possible measures include mandating the integration of green spaces in all new real estate projects, creating certifications for "biophilic buildings" that come with tax credits, and providing funding for the revegetation of existing structures. Moreover, urban planning schemes could incorporate quotas for green spaces to ensure equitable distribution of nature's benefits across the metropolitan area.

In addition, public policies should be complemented by awareness-raising campaigns and educational programs aimed at informing both citizens and decision-makers about the benefits of biophilic design. Raising public awareness about the positive effects on health, biodiversity, and climate resilience could accelerate the adoption of biophilic practices on a broader scale.

In this context, the initiative taken by the Senegalese government during World Tree Day, August 4, 2024, serves as a noteworthy example. The government initiated a nationwide reforestation campaign, emphasizing a hands-on, educational approach. Institutional authorities actively participated by planting trees in their respective localities, setting a powerful example of commitment to environmental stewardship. This initiative is one among many that should be sustained and expanded as part of a broader effort to transform Senegal's urban landscapes through biophilic design.

Such initiatives not only have the potential to enhance the urban environment but also to foster a culture of environmental responsibility and biophilic thinking among the population. This cultural shift is essential for the long-term success of biophilic design in Senegal and could serve as a model for other African nations facing similar urbanization challenges.

6 Conclusions

This chapter explores the potential of biophilic design in creating resilient, ecologically sustainable cities, with a particular focus on Dakar on its regional and metropolitan scales. The finding is that biophilic design offers a promising approach to meeting the challenges of climate change and rapid urbanization, by promoting the well-being of inhabitants, improving the energy efficiency of buildings, and strengthening the resilience of communities to natural disasters. The potential for successful integration of biophilic design into urban planning is illustrated by a series of concrete examples from Dakar, demonstrating its applicability in a variety of African contexts. However, further research and supportive policies are needed to encourage wider and effective adoption of this innovative approach.

Concerted action on the part of policy-makers, urban planners, and practitioners appears to be an imperative for integrating biophilic design into urban planning policies and development projects. By investing in biophilic initiatives such as urban parks, green roofs, and vertical gardens, Senegalese, and indeed African, cities can create healthier, more resilient, and more livable urban environments for all their inhabitants.

Ultimately, by adopting a nature-based approach to urban planning, thriving cities in harmony with their natural environment, offering an optimal living environment for current and future generations, can be shaped. En définitive, en adoptant une approche basée sur la nature dans la planification urbaine, des villes prospères en harmonie avec leur environnement naturel, offrant ainsi un cadre de vie optimal pour les générations actuelles et futures peuvent être façonnées.

References

1. Adivet (2022) Fonctions et bénéfices environnementaux. Article disponible sur: https://www.adivet.net/fonctions-et-benefices-environnementaux
2. AfDB (2022) Perspectives économiques en Afrique 2022: Soutenir la résilience climatique et une transition énergétique juste en Afrique. Rapport disponible en ligne. https://www.afdb.org/sites/default/files/documents/publications/perspectives_economiques_en_afrique_2022_web.pdf
3. Architecture Écologique (2021) "La conception biophilique dans l'architecture: principes et caractéristiques." Architecture Écologique. Disponible sur: https://architectureecologique.fr/la-conception-biophilique-dans-larchitecture-principes-et-caracteristiques/
4. Beatley T (2011) Biophilic cities: integrating nature into urban design and planning. Island Press/Center for Resource Economics, p 192
5. Berndtsson JC (2010) Green roof performance towards management of runoff water quantity and quality: a review. Ecol Eng 36(4):351–360. https://www.sciencedirect.com/science/article/abs/pii/S0925857410000029?via%3Dihub
6. Bowler DE, Buyung-Ali L, Knight TM, Pullin AS (2010) Urban greening to cool towns and cities: a systematic review of the empirical evidence. Landsc Urban Plan 97(3). https://www.sciencedirect.com/science/article/abs/pii/S0169204610001234
7. Constructive Voices (2023) Comment mettre en œuvre des solutions urbaines fondées sur la nature. Disponible sur: https://constructive-voices.com/fr/comment-mettre-en-oeuvre-des-solutions-urbaines-fondees-sur-la-nature/
8. DEFCCS (2014) Projet de réhabilitation du Parc Forestier et Zoologique de Hann au Sénégal. Direction des Eaux et Forêts, Chasses et de la Conservation des Sols, Sénégal, p 42. file:///Users/user/Downloads/Caracterisation_de_la_flore_ligneuse_du_Parc_Fores-1.pdf
9. Équilibre des énergies (2020) Construire en bois: performance environnementale et bien-être: https://www.equilibredesenergies.org/18-08-2020-construire-en-bois-performance-environnementale-et-bien-etre/
10. Gill SE, Handley JF, Ennos AR, Pauleit S (2007) Adapting cities for climate change: the role of the green infrastructure. Built Environ 33(1):115–133. https://www.ingentaconnect.com/content/alex/benv/2007/00000033/00000001/art00008
11. Heerwagen J, Hase B (2001) Building biophilia: connecting people to nature. Environmental design and construction. March–April 30–34
12. Idrissa I, Morou B, Abdourhamane H, Karim S, Abdourhamane T, Djibo I, Mahamane A (2020) Diversité floristique et structure démographique des peuplements ligneux des parcours naturels sahéliens du Sud-Est du Niger: Cas de l'enclave pastorale «Dadaria» (Mainé-Soroa, Diffa). Int J Biol Chem Sci 14(3):706–721. https://www.ajol.info/index.php/ijbcs/article/view/196831
13. Kaplan S (1995) The restorative benefits of nature: toward an integrative framework. J Environ Psychol 15(3):169–182. https://www.ajol.info/index.php/ijbcs/article/view/196831
14. Kellert SR (2008) Dimensions, elements, and attributes of biophilic design. In: Kellert SR, Heerwagen J, Mador M (eds) Biophilic design: the theory, science, and practice of bringing buildings to life. Wiley, London
15. Kellert S, Heerwagen J (2009) Nature and healing: the science, 'theory, and promise of biophilic design'. In: McDonnell MJ, Hahs AK, Breuste JH (eds) Sustainable architecture for health. Ecology of cities and towns: a comparative approach. Cambridge University Press, Cambridge. ISBN: 978-0521850940
16. Lib.Space. (2018). La conception biophilique : bénéfices et mise en œuvre dans les espaces de travail. https://www.lib.space/fr/blog/la-conception-biophilique-benefices-et-mise-en-oeuvre-dans-les-espaces-de-travail
17. Living Building Challenge. (2013). International Living Future Institute. (2013). Living Building Challenge 3.0 Standard. https://living-future.org/wp-content/uploads/2022/05/Living-Building-Challenge-3.0-Standard.pdf

18. Ministère de l'Urbanisme, du Logement et de l'Hygiène publique du Sénégal (2020) "Projet d'aménagement de la Corniche Ouest de Dakar: Le président Macky Sall satisfait des avancées." Disponible sur: https://urbanisme.gouv.sn/actualites/projet-d%E2%80%99am%C3%A9nagement-de-la-corniche-ouest-de-dakar-le-pr%C3%A9sident-macky-sall-satisfait-des
19. Mohamed S, Ewa B-A (2014) Le concept de «corridors écologiques» en milieu urbain: enjeux et contraintes d'une approche de requalification environnementale. Méditerranée [Online], 123 | 2014, Online since 01 October 2017, connection on 04 June 2024. https://journals.openedition.org/mediterranee/7537
20. Ndiaye O, Diallo A, Matty F, Thiaw A, Fall RD, Guisse A (2012) Caractérisation des sols de la zone des Niayes de Pikine et de Saint Louis (Sénégal). Int J Biol Chem Sci 6(1):519–528. https://www.ajol.info/index.php/ijbcs/article/view/80522
21. Rouchiche S (FAO) (2001) La foresterie urbaine et périurbaine en Afrique. Une étude de cas sur le Sahel (Dakar, Niamey, Nouakchott et Ouagadougou). Disponible sur: https://www.fao.org/4/x3994f/X3994F01.htm
22. Sylvain G (2017) «Arrangements locaux autour de trois fronts écologiques patrimonialisés en Afrique du Sud, Argentine et Chili». Bulletin de l'association de géographes français [En ligne], 94–2 | 2017, mis en ligne le 21 juillet 2018, consulté le 04 juin 2024. http://journals.openedition.org/bagf/1272; https://journals.openedition.org/bagf/1272
23. Tan PY, Wang J, Sia A (2013) Perspectives on five decades of the urban greening of Singapore. Cities 32:24–32
24. Terrapin Bright Green (2014) "14 Patterns of biophilic design: improving health and well-being in the built environment. "Reference: Browning WD, Ryan CO, Clancy JO (2014) 14 Patterns of biophilic design. Terrapin Bright Green llc, New York. https://www.terrapinbrightgreen.com/reports/14-patterns/
25. Thomson, G., & Newman, P. (2021). Green infrastructure and biophilic urbanism as tools for integrating resource efficient and ecological cities. *Urban Planning, 6*(1), 75–88. https://www.cogitatiopress.com/urbanplanning/article/view/3633
26. Ulrich RS, Simons RF, Losito BD, Fiorito E, Miles MA, Zelson M (1991) Stress recovery during exposure to natural and urban environments. J Environ Psychol 11(3):201–230. https://www.sciencedirect.com/science/article/abs/pii/S0272494405801847
27. United Nations (2018) "World urbanization prospects: the 2018 revision." Department of Economic and Social Affairs, Population Division. https://population.un.org/wup/Publications/Files/WUP2018-Report.pdf
28. Vroh BTA, Tiebre MS, N'guessan KE (2014) Diversité végétale urbaine et estimation du stock de carbone: cas de la commune du Plateau Abidjan, Côte d'Ivoire. Afr Sci 10(3):329–340. https://www.researchgate.net/publication/289672609_Diversite_vegetale_urbaine_et_estimation_du_stock_de_carbone_cas_de_la_commune_du_Plateau_Abidjan_Cote_d'Ivoire/link/571a3a7408ae6eb94d0c7538/download
29. Wilson EO (1984) Biopbilia: the human bond with other species. Harvard University Press, Cambridge

Integrating Blue Zone Principles into Healthy City Design

Ali Cheshmehzangi, Zaheer Allam, Xin Wang, and Zhaohui Su

Abstract This chapter provides a thorough framework for creating cities that support sustainability, longevity, and good health by examining the relationship between climate-resilient urbanism and Blue Zone principles. Using knowledge from Blue Zones—areas with a high life expectancy—it looks at how urban planning can encourage regular exercise, social interaction, a balanced diet, and a sense of purpose in life. The chapter also highlights how policies can support climate resilience and create environments that support these health-promoting behaviours. It draws attention to how technology, neighbourhood-based strategies, and green infrastructure can influence how people live in cities in the future. Cities can address the increasing challenges posed by climate change while improving the well-being of their citizens by combining climate-responsive strategies with health-focused design. In light of the urban challenges of the twenty-first century, this synthesis offers a guide for cities looking to strike a balance between sustainability and public health.

Keywords Blue zones · Climate resilience · Urban health · Sustainable cities · Urban policy · Green infrastructure · Longevity

1 Introduction

The rising emphasis on urban resilience and health-focused urbanism has spotlighted a compelling model for fostering well-being: the Blue Zones [1]. These areas—Sardinia, Italy; Nicoya, Costa Rica; Ikaria, Greece; Okinawa, Japan; and Loma Linda, California—are well known for their high numbers of centenarians and their unique,

A. Cheshmehzangi (✉) · Z. Allam
School of Architecture, Design and Planning, The University of Queensland, Brisbane, Australia
e-mail: A.Chesh@uq.edu.au

X. Wang
College of Architecture and Urban Planning, Tongji University, Shanghai, China

Z. Su
School of Public Health, Southeast University, Nanjing, China

A. Cheshmehzangi et al. (eds.), *Designing Healthy Cities*, Urban Sustainability,
https://doi.org/10.1007/978-981-96-5538-0_5

health-promoting ways of living [2]. People who live in the Blue Zone have long life expectancies, few chronic illnesses, and generally good physical, mental, and social health [3]. These areas are unique because they combine behavioural, cultural, and environmental elements that support longevity and healthy aging. In order to build healthier, more sustainable cities, this chapter examines how the ideas of Blue Zones can be incorporated into urban planning.

The Blue Zone lifestyles' compatibility with urban planning ideas provides a revolutionary method of city planning [4]. Policymakers and urban planners have the chance to incorporate important Blue Zone ideas into city infrastructures that promote stress management, social interaction, active living, and plant-based diets [5]. In addition to increasing the healthy lifespan of urban dwellers, such initiatives could support climate-resilient urbanism, which is essential for addressing today's pressing global issues.

This chapter explores five important ways that incorporating Blue Zone principles can result in the development of more resilient, healthy cities. After discussing the characteristics that make up Blue Zones [6], we look at the principles of healthy city planning. Next, we examine how the two intersect, highlighting real-world uses and possible effects on urban populations [7]. Lastly, we go over the difficulties and solutions for incorporating these ideas into contemporary city planning frameworks.

1.1 Understanding Blue Zones: Characteristics and Health Outcomes

Dan Buettner and his research team came up with the idea of "blue zones," which are places where people live longer, healthier lives. In addition to having lower rates of chronic illnesses like diabetes, heart disease, and cancer, residents of these areas frequently live to be over 100. The defining traits of Blue Zones and the elements that lead to the remarkable longevity and health of their inhabitants are thoroughly examined in this section.

Key characteristics of Blue Zones [7–16] include (see Fig. 1):

- **Diet:** Inhabitants of Blue Zones consume primarily plant-based diets, rich in whole grains, legumes, vegetables, and fruits, with limited meat consumption. The Mediterranean diet, for instance, is prevalent in Sardinia and Ikaria, while Okinawans have a diet centered around sweet potatoes and tofu.
- **Physical Activity:** Blue Zone residents do not engage in structured exercise but instead maintain high levels of natural physical activity, often linked to daily tasks such as farming, gardening, and walking.
- **Social Connectivity:** A strong sense of community and social integration is a hallmark of Blue Zones. People in these areas often have close-knit social circles that provide emotional support, reduce stress, and encourage healthy behaviors.
- **Purpose and Mental Well-being:** A key aspect of longevity in Blue Zones is having a sense of purpose, often referred to as "Ikigai" in Japan or "Plan de Vida"

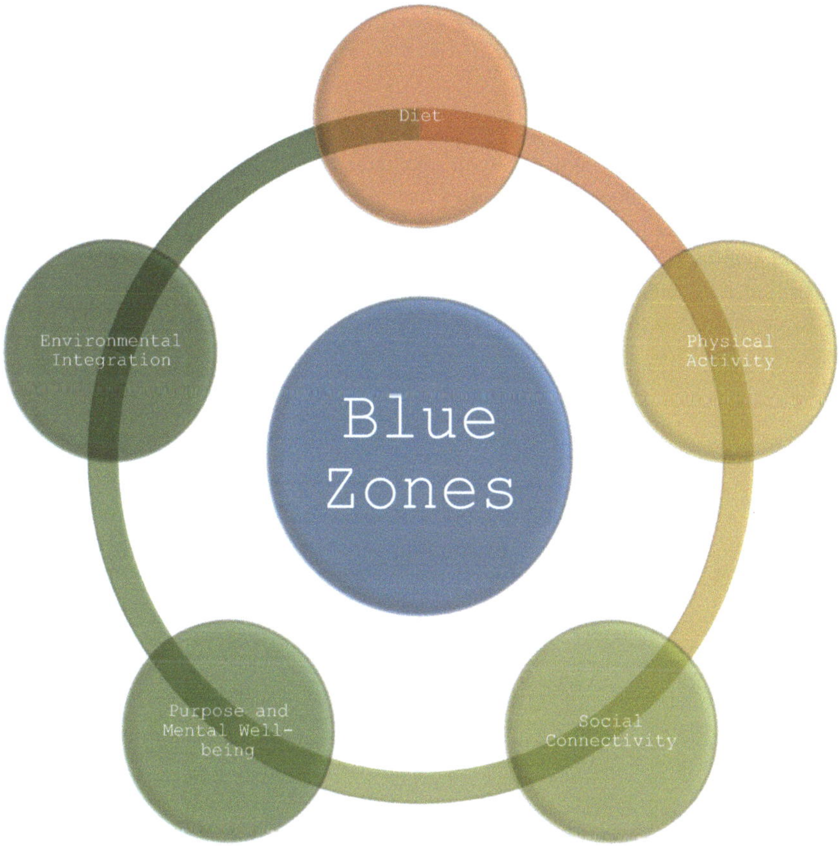

Fig. 1 Summary of main characteristics of Blue Zones

in Costa Rica. This purpose-driven life is linked to mental well-being, resilience, and a positive outlook on aging.

- **Environmental Integration:** Blue Zone communities are deeply connected to their natural environments. The landscapes encourage walking, outdoor activities, and communal interaction.

The combination of these factors forms the foundation of their longevity, offering a replicable blueprint for other regions and cities aiming to enhance public health and well-being.

1.2 Principles of Healthy City Design and Climate-Resilient Urbanism

Urban design must change to incorporate the concepts of healthy living and climate resilience as cities deal with the twin challenges of promoting public health and adjusting to climate change [17]. Cities that offer settings that support social interaction, mental health, and physical activity are considered healthy. The fundamental ideas of healthy city planning are discussed in this section, along with the significance of resilience against climate-related hazards.

Key principles of healthy cities [18–26] include (see Fig. 2):

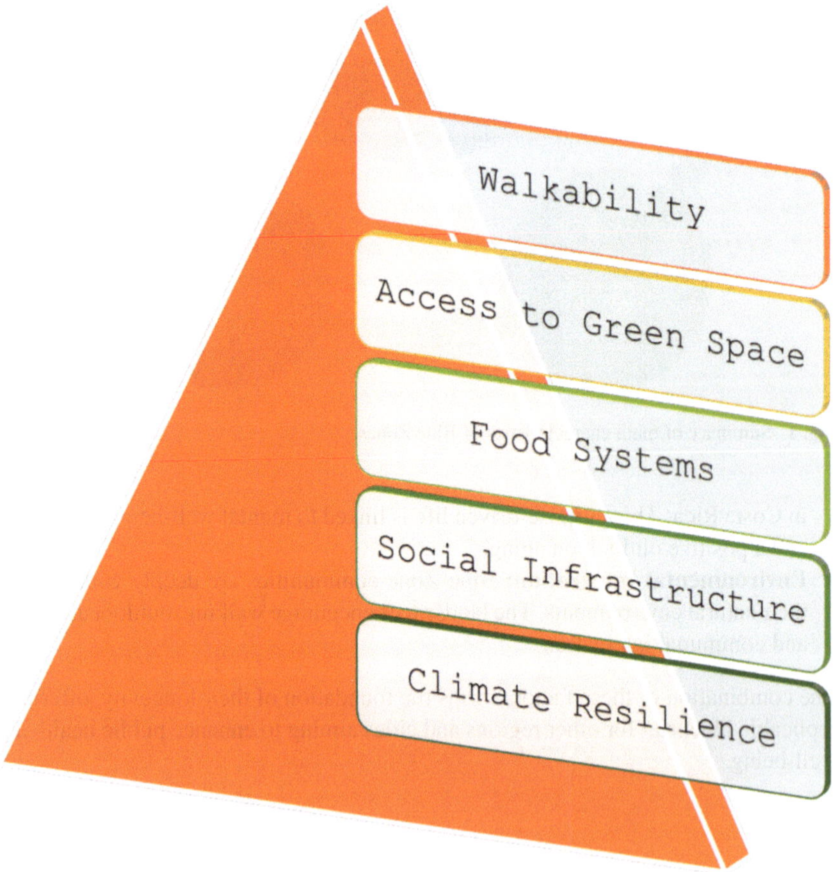

Fig. 2 Summary of key principles of healthy cities

- **Walkability:** Designing cities with pedestrians in mind encourages walking as a form of daily transportation. Walkable cities promote physical health by reducing car dependency and increasing opportunities for exercise.
- **Access to Green Spaces:** Urban parks, gardens, and other green spaces are vital for reducing air pollution, mitigating heat islands, and offering areas for recreation and social interaction. Studies have shown that access to nature improves mental health and stress levels.
- **Food Systems:** Urban food environments should promote access to healthy, affordable food. This includes creating urban farms, community gardens, and markets that provide fresh, local produce, drawing inspiration from the plant-based diets of Blue Zones.
- **Social Infrastructure:** Public spaces and community centers are integral to fostering social interaction and reducing social isolation, which is a significant factor in health outcomes.
- **Climate Resilience:** Designing cities to withstand climate change involves considering flood management, energy efficiency, and reducing carbon emissions. Green infrastructure—like permeable pavements, rooftop gardens, and stormwater management systems—plays a pivotal role in enhancing urban resilience.

Together, these elements create environments that support active, engaged lifestyles while protecting residents from the impacts of a warming climate.

2 Applying Blue Zone Principles in Sustainable Urban Design

A thorough grasp of how physical environments influence daily behaviours and, in turn, long-term health outcomes is necessary for the incorporation of Blue Zone principles into urban design [27]. Blue Zones, which are defined by the organic integration of health-promoting behaviours into everyday life, provide important information about how cities can be planned to encourage social interaction, physical activity, and overall well-being. When applying these ideas to urban settings, it's critical to design spaces that inherently promote longevity-promoting behaviours like consistent exercise, social interaction, a balanced diet, and a feeling of purpose.

The significance of creating urban settings that encourage regular, organic physical activity is one of the main lessons to be learned from Blue Zones. In contrast to conventional exercise regimens, which are frequently planned and distinct from daily activities, people who live in the Blue Zone incorporate movement into their daily lives. This can be replicated in urban design by emphasizing walkability and minimizing reliance on automobiles. The layout of cities should be such that residential areas can walk to basic services like supermarkets, schools, and medical facilities. Active transportation is also promoted by the construction of bike infrastructure and safe, accessible pedestrian walkways. It is possible to create public areas that encourage incidental physical activity, making movement a natural part of socializing, leisure, and commuting.

In addition to making mobility easier, urban settings need to promote social connectivity, which is essential to the Blue Zone way of life [28]. Residents' mental health and general well-being are greatly enhanced by the robust social networks that offer emotional support in Blue Zones. In an urban setting, this means creating areas that promote social cohesiveness and community involvement. Communities should be organized around public areas where people can congregate and socialize, like parks, plazas, and community centers. Public areas ought to be multipurpose places for socializing, cultural events, and recreation. Human-scale urbanism, which stresses the development of settings that feel approachable, friendly, and supportive of impromptu social interactions, can be incorporated into the design of these areas. Urban dwellers' quality of life is improved by a well-designed public realm, which fosters mental and social health by lowering stress and fostering a sense of community.

The prevalence of plant-based diets, which have been repeatedly linked to lower rates of chronic disease and longer life expectancy, is another important characteristic of Blue Zones [29]. By increasing access to local, fresh produce and decreasing the occurrence of food deserts—urban regions with limited access to reasonably priced, healthful food options—urban design can encourage people to eat healthier. Policy interventions, like zoning laws that support the establishment of supermarkets and farmers' markets in underprivileged areas, can help achieve this. Cities can also encourage rooftop or vertical farming projects and provide space for community gardens as ways to advance urban agriculture [30]. In addition to increasing access to nutritious food, these interventions strengthen the bond between locals and their food systems, which is central to the Blue Zone philosophy.

Urban design can also promote the idea of purposeful living, which has been found to be a major factor in longevity in Blue Zones. Better mental health, resilience, and life satisfaction are associated with a sense of purpose, which is frequently demonstrated through meaningful work, volunteer work, and close family or community ties. Purposeful urban environments offer chances for community involvement, creative expression, and lifelong learning. Libraries, cultural centers, and makerspaces are examples of educational and skill-building spaces that city planners can integrate into urban neighbourhoods. Additionally, programs that encourage intergenerational relationships, volunteerism, and mentoring can assist people in discovering and fostering their sense of purpose within the urban fabric. Providing physical spaces is only one aspect of purpose-driven urbanism; another is cultivating a culture that values and incorporates community contributions and personal development into daily life.

Last but not least, a key component of Blue Zones is the close bond that exists between humans and their natural surroundings [31, 32]. Biophilic design, an approach that aims to incorporate natural elements into urban spaces, can help cities foster this connection, which is closely linked to both physical and mental well-being. Cities can incorporate green infrastructure, such as parks, greenways, urban forests, and waterways, to give their citizens access to calming, accessible natural spaces [33]. It has been demonstrated that having nature in urban areas lowers stress, elevates mood, and promotes physical activity [34]. By reducing the urban heat island effect, controlling stormwater, and improving air quality, green spaces not only benefit individuals [35] but also improve the ecological health of cities and increase their overall resilience to climate change.

In summary, applying Blue Zone concepts to urban planning entails a comprehensive strategy that gives equal weight to community resilience and individual well-being [36–38]. Urban planners can build cities that not only improve the health and longevity of their citizens but also help create sustainable, climate-resilient futures by designing spaces that naturally encourage movement, social interaction, healthy eating, purposeful living, and a connection to nature. This integrated approach to urbanism is a crucial tactic for creating the healthy cities of the future because it balances the needs of people with more general environmental and social objectives.

3 The Role of Policy in Promoting Healthy, Resilient Cities

It takes more than just planners and architects to successfully incorporate Blue Zone principles into urban design; a strong legislative framework that encourages and institutionalizes resilient, health-promoting urban development is also necessary. By ensuring that social programs, governance structures, and urban infrastructure are all in line with these objectives, policy serves as the cornerstone for developing environments that promote longevity, sustainability, and well-being. A carefully thought-out policy environment can improve resilience to the impending challenges posed by climate change, control urban growth to prioritize livability, and encourage the adoption of healthy behaviors. This section explores how policy plays a critical role in forming climate-resilient and healthy cities, emphasizing how policy interventions can produce urban ecosystems that replicate the health outcomes observed in Blue Zones.

Land-use and zoning regulations are two fundamental areas where policy is essential. The physical layout of cities is directly influenced by the zoning laws that regulate urban development, which establish the locations of residential, commercial, and recreational areas. Zoning regulations must give mixed-use projects that encourage walkability and lessen dependency on cars top priority in cities hoping to implement Blue Zone principles. Residents may choose to walk or cycle instead of drive if zoning encourages compact, dense neighbourhoods with easy access to necessary services. Policies that guarantee public access to parks, leisure centers, and green areas are also crucial for creating settings where regular exercise is incorporated into everyday life. Zoning regulations help create urban environments that naturally encourage the kinds of health-promoting behaviours associated with Blue Zones by controlling how close housing is to food outlets, green spaces, and social hubs.

Policies that facilitate the availability of reasonably priced, healthful food are also essential for extending life expectancy and lowering the prevalence of chronic diseases linked to diet. To improve food systems, governments can implement a variety of policy interventions, such as regulating the location of grocery stores in underserved areas or offering incentives for local food production. Urban agricultural policies, for example, can support the growth of local farms and community gardens, giving locals access to wholesome, fresh food while also fostering environmental sustainability. By providing tax breaks or grants to supermarkets and farmers' markets that open in underprivileged or low-income areas, local governments can further their efforts to eradicate food deserts. The dietary patterns that support longevity in Blue Zones can be directly addressed by policies that establish economic conditions that make healthy food affordable and accessible. Additionally, by aligning urban food environments with Blue Zone principles, policies that support the consumption of plant-based foods through public education campaigns or government-backed dietary guidelines can encourage healthier eating habits throughout the entire city.

Another area where policy intervention can have a major impact on health is social infrastructure. Strong social networks that support well-being and mental health can only be fostered by policies that support the creation of public areas intended for community interaction. Social cohesion in Blue Zones is essential for residents' longevity because it lowers stress and fosters emotional support. By allocating funds for the development and upkeep of community centers, parks, and public plazas, governments can promote social connectivity. Policies that place a high priority on incorporating common areas into new construction guarantee that inhabitants have access to settings where they can interact with one another and form bonds. Furthermore, laws that promote inclusive zoning and affordable housing help avoid social segregation by guaranteeing that different groups can coexist and communicate in public areas. Governments can encourage the sense of community that is so vital to well-being by implementing urban policies that encourage inclusivity and social interaction.

The effectiveness of urban environments that promote health is greatly influenced by public health policies. By launching public health initiatives that inform the public about the advantages of embracing Blue Zone practices—such as greater physical activity, better eating habits, and stress-reduction techniques—governments can promote healthy cities. By focusing interventions on populations most at risk of social isolation or chronic illness, public health policies can also take the initiative to address health disparities. Cities can, for instance, put in place initiatives that support active transportation, like bike-sharing programs, public exercise classes, or "safe routes" initiatives that encourage kids to ride their bikes or walk to school. Furthermore, in order to lower the prevalence of non-communicable diseases, governments can collaborate with healthcare providers to create wellness initiatives that provide incentives for preventive care. In order to ensure that longevity and well-being are given priority in urban life, public health initiatives that are in line with the tenets of Blue Zones can aid in institutionalizing these behaviours across entire populations.

3.1 Climate-Resilience and Facing Climate Change Impacts on Cities

Promoting climate resilience, a growing concern for cities dealing with the effects of climate change, is another crucial aspect of policy [39, 40]. The health and well-being of urban populations are at risk from environmental shocks like heatwaves, rising sea levels, and extreme weather events, which resilience policies make sure urban infrastructure can withstand [41]. In order to build cities that not only promote human longevity but also protect the environment that sustains human life, climate-resilient urban design must coexist with health-promoting policies. Cities can better manage stormwater, lessen the urban heat island effect, and enhance air quality by implementing policies that support the development of green infrastructure, such as rain gardens, green roofs, and permeable pavements. Similar to this, policies that promote the use of energy-efficient building materials and the switch to renewable energy sources help to reduce urban pollution and build climate resilience, both of which have a positive impact on public health.

Long-term planning calls for the integration of resilience strategies into urban governance, which means that all levels of government must be accountable for both the more general environmental sustainability issues and the more pressing health needs. The intricacy of building resilient, healthy cities is best addressed by policy frameworks that promote cross-sectoral cooperation. Policy approaches must encourage interdisciplinary solutions in order to reflect the interconnectedness of infrastructure development, environmental sustainability, and urban health. In order to develop integrated policies that simultaneously address public health and climate resilience, governments can help to establish collaborations between urban planners, public health specialists, environmental scientists, and community organizations.

To sum up, policy is essential to creating cities that support Blue Zone ideals and the more general objectives of climate resilience. Governments have the ability to design urban environments that support residents' long, healthy, and satisfying lives through zoning laws, food access programs, public health campaigns, and resilience tactics. A way to build cities that not only lessen the effects of climate change but also improve the quality of life for their citizens is to align urban policy with the lifestyle choices and behaviours that define Blue Zones. Policy's role in integrating resilience and health into urban planning will be crucial as cities around the world expand and change, helping to create liveable, sustainable spaces that learn from the longest-living communities in the world.

3.2 Future Prospects: Innovating for Healthier, Resilient Urban Living

The combination of climate-resilient urbanism and Blue Zone principles offers cities a strong chance to develop in ways that support environmental sustainability and public health. Cities are confronted with previously unheard-of difficulties in managing resources, public health, and climate change as urban populations continue to rise. But these difficulties also present a singular chance to reconsider conventional urban models and integrate approaches that put resilience, longevity, and well-being first. Urban living's future depends on combining climate resilience with health-conscious design, utilizing data-driven policymaking [42], community-driven solutions, and technology breakthroughs to build more liveable, sustainable spaces.

The use of technology in urban health and resilience is one of the most exciting areas for future innovation. Smart cities have enormous potential to create healthier urban environments by using digital infrastructure to monitor, assess, and manage urban systems. Real-time data collection on energy, water, and air quality, for example, can help guide more effective resource management and guarantee that cities are sensitive to environmental threats. On days with poor air quality, city officials can implement measures like reducing traffic or improving public transportation because sensors positioned throughout urban areas can detect changes in pollution levels. These data-driven approaches align with Blue Zone principles by promoting environments that naturally support healthier behaviours—clean air, green spaces, and sustainable transit options that reduce stress and promote physical activity.

Beyond managing environmental systems, technology also offers potential for fostering healthier lifestyles among urban populations. With the rise of wearable technology and health monitoring applications, people can now track their levels of activity, keep an eye on their diets, and instantly manage stress. To promote physical activity, these technologies can be incorporated into the city's infrastructure [29]. Cities can, for instance, implement digital platforms that offer up-to-date information on routes for bicyclists and pedestrians, incorporate health data with public transportation, or gamify physical activity by rewarding active commuting. These developments are especially pertinent to urbanism influenced by the Blue Zone, where physical activity is integrated into daily life rather than existing as a standalone activity. By using technology to make healthy behaviours more accessible and rewarding, cities can better align with the health-promoting principles observed in Blue Zones.

The development of resilient, healthy cities in the future will depend on community-driven strategies in addition to technology advancements. Future city design must continue to prioritize the social cohesion and human connections that are essential to Blue Zones, even though smart technologies can improve urban living. In order to make sure that city design represents the needs and values of its citizens, urban planners must concentrate on enabling communities to actively participate in the development of their surroundings. Participatory urban planning techniques that involve local communities in decision-making can help achieve this. Cities can build more inclusive, equitable, and population-responsive environments by integrating citizen input on topics like the location of green spaces, the layout of public transportation systems, and the creation of community health programs.

Additionally, community-based programs offer chances to cultivate social ties, which are essential for mental and emotional health. Cities can, for example, encourage shared housing models, intergenerational living arrangements, or community gardens that foster cooperation and mutual support among residents. These programs can replicate the close-knit social bonds that define Blue Zone communities and lessen social isolation, a problem that is becoming more prevalent in many urban areas. Future cities must prioritize developing social infrastructure that promotes trust, cooperation, and a feeling of community in addition to developing physical infrastructure [43]. By ensuring that social well-being continues to be a fundamental element of urban resilience and health, community-driven urbanism serves as a complement to the technological developments that are reshaping contemporary cities.

The creation of flexible and adaptive urban areas that can react to shifting environmental and health demands is another crucial factor for future cities [44]. Cities need to be built with the ability to endure these disruptions while still promoting the health and well-being of their citizens [45], as climate change is predicted to bring more frequent and severe weather events. Resilience must be given top priority in future urban development by implementing design elements like drought-tolerant landscaping, heat-reflective building materials, and flood-resistant infrastructure. Cities must be flexible in terms of governance and policy in addition to being physically resilient [46]. This will allow them to react quickly to new environmental threats or pandemics, among other public health emergencies. This requires forward-thinking urban governance that anticipates risks, integrates health and climate data into decision-making, and ensures that cities can evolve in response to shifting conditions.

4 Concluding Remarks

The development of resilient, healthy cities in the future will also be greatly influenced by innovations in green infrastructure. In addition to being crucial for climate resilience, green roofs, urban forests, and water-sensitive design techniques also improve the physical and emotional well-being of city dwellers [47, 48]. The principles of biophilic design, which highlight the value of human connection to the natural world, are in line with the incorporation of nature into the built environment. Access to green spaces can lower stress, elevate mood, and promote physical activity, all of which lead to longer, healthier lives, as demonstrated in Blue Zone communities [49]. Future cities will need to investigate novel approaches to integrating nature into urban settings, such as turning unused areas into green corridors, establishing vertical gardens, and coming up with stormwater management strategies based on nature. By providing a model for sustainable urban living, these advancements in green infrastructure will improve cities' ecological resilience and the health of their citizens.

Last but not least, data-driven policy strategies will be crucial to guaranteeing that future cities continue to support resilience and health objectives. To monitor the success of programs aimed at fostering resilience and promoting health, governments and urban planners will need to use data. This could involve evaluating the social benefits of community-driven planning projects, tracking improvements in air quality brought about by green infrastructure, or using data to gauge how urban design affects levels of physical activity. In order for cities to implement targeted interventions for vulnerable populations, identify gaps in health outcomes, and make sure that urban environments are continuously optimized to promote well-being and climate resilience, data will be essential. Future cities can be more responsive and adaptable by adopting data-driven policy, employing evidence-based tactics to create urban environments that are healthier and more sustainable.

In summary, Blue Zone principles provide a useful framework for directing this evolution, and the future of urban living lies at the nexus of resilience and health [50, 51]. There has never been a more pressing need for creative, progressive urban design as cities deal with the effects of population growth, climate change, and public health issues. The development of cities that support longevity, well-being, and climate resilience will heavily rely on technology, community involvement, flexible urban areas, and green infrastructure. The ongoing development of data-driven policy approaches that guarantee urban environments are responsive to the needs of their inhabitants and the challenges of the future, however, will be necessary for the success of these innovations. Urban planners can create environments that reflect the health-promoting qualities of Blue Zones by incorporating these components into the planning and administration of future cities. This will provide a resilient and sustainable model for urban living in the twenty-first century and beyond.

References

1. Fedorowicz M, Schilling J, Bramhall E, Bieretz B, Su Y, Brown S (2020) Leveraging the built environment for health equity. Urban Institute. Available from: https://www.urban.org/sites/default/files/publication/102557/leveraging-the-built-environment-for-health-equity.pdf
2. Albronda FT (2016) Sustainable aging as a community goal: how the community-based blue zones project strives for a healthier society in cedar falls, Iowa. Bachelor's thesis
3. Cheshmehzangi A, Zou J, Sharifi A, Zhang R, Ziafati Bafarasat A, Zhao J (eds) (2025) Designing healthy cities: integrating climate-resilient urbanism for sustainable living. Springer, Singapore
4. Buettner D (2023) The blue zones secrets for living longer: lessons from the healthiest places on earth. National Geographic Partners, LLC, Washington, DC
5. Koes E (2024) Healthy ageing through the living environment. Available from: https://repository.tudelft.nl/file/File_71784c09-3f65-4403-863a-7f91976f2860?preview=1
6. Bell S, Fleming LE, Grellier J, Kuhlmann F, Nieuwenhuijsen MJ, White MP (eds) (2021) Urban blue spaces: planning and design for water, health and well-being. Routledge, New York
7. Marston HR, Niles-Yokum K, Silva PA (2021) A Commentary on Blue Zones®: a critical review of age-friendly environments in the 21st century and beyond. Int J Environ Res Public Health 18(2):837
8. Aliberti SM, Donato A, Funk RH, Capunzo M (2024) A narrative review exploring the similarities between Cilento and the already defined "blue zones" in terms of environment, nutrition, and lifestyle: can Cilento be considered an undefined "blue zone"? Nutrients 16(5):729
9. Beer GO (2022) Blue zones: unlocking key themes in the centenarian's life. Available from: https://digitalcommons.murraystate.edu/honorstheses/115/

10. Buettner D, Skemp S (2016) Blue zones: lessons from the world's longest lived. Am J Lifestyle Med 10(5):318–321
11. Herbert C, House M, Dietzman R, Climstein M, Furness J, Kemp-Smith K (2022) Blue zones: centenarian modes of physical activity: a scoping review. J Popul Ageing 2022:1–37
12. Hitchcott PK, Fastame MC, Penna MP (2018) More to blue zones than long life: positive psychological characteristics. Health Risk Soc 20(3–4):163–181
13. Kreouzi M, Theodorakis N, Constantinou C (2024) Lessons learned from Blue Zones, lifestyle medicine pillars and beyond: an update on the contributions of behavior and genetics to wellbeing and longevity. Am J Lifestyle Med 18(6):750–765
14. Nedelcu E (2018) The impact of social relationships over health and longevity. The "Blue Zones" case. Roman Rev Soc Sci 14:1
15. Poulain M, Herm A, Pes G (2013) The Blue Zones: areas of exceptional longevity around the world. Vienna Yearb Popul Res 2013:87–108
16. Santacroce L, Bottalico L, Charitos IA, Haxhirexha K, Topi S, Jirillo E (2024) Healthy diets and lifestyles in the world: Mediterranean and Blue Zone people live longer Special focus on gut microbiota and some food components. Endocrine Metab Immune Disord Drug Targets 24(15):1774–1784
17. Cushing DF, Miller E (2019) Creating great places: evidence-based urban design for health and wellbeing. Routledge, New York
18. Bailey JE, Mirvis DM, Key CM, Kyte RL, McCord MJ (2007) HealthCARE principles: a model for healthy city collaboratives. Prev Chronic Dis 4(2):1
19. Barton H, Grant M (2013) Urban planning for healthy cities: a review of the progress of the European Healthy Cities Programme. J Urban Health 90:129–141
20. Davies JK, Kelly M (2014) Healthy cities: research and practice. Routledge, New York
21. Duhl LJ, Sanchez AK, World Health Organization (1999) Healthy cities and the city planning process: a background document on links between health and urban planning (no. EUR/ICP/CHDV 03 04 03). WHO Regional Office for Europe, Copenhagen
22. Heritage Z, Dooris M (2009) Community participation and empowerment in healthy cities. Health Promot Int 24(Suppl_1):i45–i55
23. Kim JH (2012) Principles for evaluating healthy cities. Korean J Health Educ Promot 29(4):67–81
24. Lawrence R, Fudge C (2007) 16 healthy cities: key principles for professional practice. Public Health Soc Context Act 2007:180
25. Ramaswami A, Russell AG, Culligan PJ, Sharma KR, Kumar E (2016) Meta-principles for developing smart, sustainable, and healthy cities. Science 352(6288):940–943
26. Tsouros AD (1995) The WHO healthy cities project: state of the art and future plans. Health Promot Int 10(2):133–141
27. Engineer A, Gualano RJ, Crocker RL, Smith JL, Maizes V, Weil A, Sternberg EM (2021) An integrative health framework for wellbeing in the built environment. Build Environ 205:108253
28. Najafi P, Mohammadi M (2024) Redefining age-friendly neighbourhoods: translating the promises of blue zones for contemporary urban environments. Int J Environ Res Public Health 21(3):365
29. Cheshmehzangi A (2021) Urban health and sustaining peace during and after the COVID-19 pandemic. J Infect Dis Prev Med 9(3):1000219
30. Khan MM, Akram MT, Janke R, Qadri RWK, Al-Sadi AM, Farooque AA (2020) Urban horticulture for food secure cities through and beyond COVID-19. Sustainability 12(22):9592
31. Boyd F, Allen C, Robinson JM, Redvers N (2024) The past, present, and future of nature and place-based interventions for human health. Landsc Res 49(1):129–145
32. Sáenz de Tejada C, Daher C, Hidalgo L, Netanyahu S, Nieuwenhuijsen M, Braubach M (2024) Urban planning, design and management approaches to building urban resilience: a rapid review of the evidence. Cities Health 8(5):932–955
33. Mancilla D, Robledo S, Esenarro D, Raymundo V, Vega V (2024) Green corridors and social connectivity with a sustainable approach in the city of Cuzco in Peru. Urban Sci 8(3):79

34. Liu Z, Xiu C, Ye C (2020) Improving urban resilience through green infrastructure: an integrated approach for connectivity conservation in the Central City of Shenyang, China. Complexity 2020(1):1653493
35. Courtney-Wolfman L (2024) Urban heat waves and adaptive capacity: How can social infrastructure help reduce vulnerability? In: Handbook of social infrastructure. Edward Elgar Publishing, Cheltenham, pp 289–311
36. Kent JL, Thompson S (2014) The three domains of urban planning for health and well-being. J Plan Lit 29(3):239–256
37. Ling TY, Chiang YC (2018) Well-being, health and urban coherence-advancing vertical greening approach toward resilience: a design practice consideration. J Clean Prod 182:187–197
38. Smith N, Georgiou M, King AC, Tieges Z, Chastin S (2022) Factors influencing usage of urban blue spaces: a systems-based approach to identify leverage points. Health Place 73:102735
39. Acuti D, Bellucci M (2020) Resilient cities and regions: planning, initiatives, and perspectives. Clim Act 2020:763–774
40. Allam Z, Jones D, Thondoo M (2020) Cities and climate change: climate policy, economic resilience and urban sustainability. Disney Electronic Content. Springer, Singapore
41. World Health Organization (2022) Urban planning, design and management approaches to building resilience—an evidence review: first report on protecting environments and health by building urban resilience. Available online at: https://iris.who.int/handle/10665/355761
42. Pourroostaei Ardakani SP, Cheshmehzangi A (2023) Big data analytics for smart transport and healthcare systems. Springer, Singapore
43. Ziafati Bafarasat A, Cheshmehzangi A, Ankowska A (2023) A set of 99 healthy city indicators for application in urban planning and design. Sustain Dev 31(3):1978–1989
44. Carter JG, Cavan G, Connelly A, Guy S, Handley J, Kazmierczak A (2015) Climate change and the city: building capacity for urban adaptation. Prog Plan 95:1–66
45. Birkmann J, Garschagen M, Kraas F, Quang N (2010) Adaptive urban governance: new challenges for the second generation of urban adaptation strategies to climate change. Sustain Sci 5:185–206
46. Javadinejad S, Dara R, Jafary F (2021) Analysis and prioritization the effective factors on increasing farmers resilience under climate change and drought. Agric Res 10:497–513
47. Cheshmehzangi A (2020) Reflection on early lessons for urban resilience and public health enhancement during the COVID-19. Health 12(10):1390
48. Cheshmehzangi A (2020) How cities cope in outbreak events? The city in need: urban resilience and city management in disruptive disease outbreak events, pp 17–39
49. World Health Organization (2017) Urban green space interventions and health: a review of impacts and effectiveness. Available from: https://apps.who.int/iris/handle/10665/366036
50. Cheshmehzangi A (2021) Urban health, sustainability, and peace in the day the world stopped. Springer, Singapore
51. Cheshmehzangi A, Sedrez M, Zhao H, Li T, Heath T, Dawodu A (eds) (2023) Resilience vs. pandemics: innovations in cities and neighbourhoods. Springer, Singapore

Vernacular Heat-Resilient Cities in Hot-Humid Climates: Exploring Solutions and Strategies for Bushehr, Iran

Maryam Roosta and Roza Vakilinezhad

Abstract The concept of "heat-resiliency" has recently gained significant attention in research, though historical communities have long adapted to environmental and climatic conditions through practical and experimental wisdom. This study aims to extract heat-resilient design strategies from the historical urban fabric of Bushehr, Iran, at both the building and community levels. Bushehr is a historical city with a hot and humid climate located in the north of the Persian Gulf. Its urban fabric spans approximately three centuries and covers 38 hectares, including over 90 valuable historical houses. To achieve the research aim, data were collected from library sources and from old and new maps and images of Bushehr's historical fabric. A framework of strategies, components, and solutions to improve heat resiliency in this historical context was developed through content analysis and logical reasoning. This framework addresses three scales: macro (morphology and general appearance of the fabric), meso (blocks and urban network), and micro (building architecture and details). Key solutions identified include compactness, the orientation of the fabric and urban network, enclosure, intersections, block size and permeability, central courtyards, semi-open spaces such as Tarameh and Shanashir, design and proportions of openings and shades, color and thermal capacity of materials, and greenery in courtyards. This research can inform the planning and design of new urban settlements, particularly in hot and humid areas, to adopt heat-resiliency policies in urban planning and architecture.

Keywords Vernacular cities · Heat-resilience · Urban cooling · Bushehr

M. Roosta (✉)
Department of Urban Planning and Design, Shiraz University, Shiraz, Iran
e-mail: m-roosta@shirazu.ac.ir

R. Vakilinezhad
Department of Architecture, Shiraz University, Shiraz, Iran

A. Cheshmehzangi et al. (eds.), *Designing Healthy Cities*, Urban Sustainability,
https://doi.org/10.1007/978-981-96-5538-0_6

1 Introduction

Rapid urbanization has significantly transformed the human living environment. The built environment has become warmer and denser, with diminished greenery and the formation of microclimates. Global warming has exacerbated this issue by increasing energy consumption for heating, ventilation, and air conditioning, raising greenhouse gas emissions [1, 2]. The building sector significantly impacts the environment, accounting for approximately 30% of greenhouse gas emissions and 40% of global energy consumption. In 2020, 40–45% of energy consumption in the European Union and about 51% of carbon emissions in China were attributed to the building sector [3].

In recent decades, the increase in energy demand and air pollution has compromised the thermal comfort of people. Thermal comfort plays a fundamental role in living, activity, and residence in indoor spaces, and it also affects the presence of people in urban spaces, economic vitality, and tourism [4, 5]. Urban design and planning are influential, with the arrangement of the built environment from micro to macro scales impacting residents' thermal comfort [6].

Heat resilience in hot climates is important, but achieving thermal comfort in hot and humid climates poses a more serious challenge. Studies show that residents in hot and humid areas have significantly lower heat resilience and a higher demand for lower temperatures compared to those in hot and dry climates [7]. High humidity alongside heat reduces the range of human thermal comfort and the ability to adapt to climatic conditions [7, 8]. Furthermore, more research on thermal comfort in hot and humid climates, especially in outdoor spaces, is still required [6].

Passive design is one of the most effective strategies for reducing energy consumption in the building sector. While cooling systems and air conditioners account for 15% of electricity consumption, significant savings can be achieved through natural ventilation methods [9, 10, 11]. Research in hot and humid climates shows that natural ventilation strategies can reduce 22% of the radiant heat received by buildings [12]. These strategies include building shape, orientation, volume distribution, and leveraging site potentials to minimize solar energy and utilize prevailing winds [11]. Extensive research has highlighted the importance of passive methods in enhancing thermal comfort in hot and humid areas, such as in Saudi Arabia [13], UAE [14], Mexico [15], Taiwan [16], Malaysia [17], and India [18]. Climate-responsive design has become essential to address the energy crisis and climatic comfort [2, 19].

The best examples of passive and climate-responsive design can be found in traditional and vernacular structures. These structures are valuable sources of inspiration for creating sustainable architecture and cities [20]. Patterns and strategies used in traditional contexts, typically obtained through human trial and error, effectively respond to local needs with limited resources. Reinterpreting them for modern architecture and urbanism is necessary [2, 21]. Vernacular buildings harmonize human behavior, culture, traditions, construction techniques, and environmental requirements [22, 23]. Studies show that applying vernacular design strategies can achieve better thermal and ventilation performance in architecture and cities [24, 25, 26,

27]. However, more localized research is needed to distinguish microclimates and different urban areas in diverse climatic and cultural settings to reach comprehensive guidelines. Yang and colleagues reviewed 55 studies on climate-responsive indigenous habitats and believe that more quantitative research is needed for future climate-compatible construction [3].

Many studies focus on reinterpreting traditional construction principles and strategies in architecture and urban planning to achieve environmental and climatic comfort. In Nepal, traditional construction principles are well adapted to the local climatic conditions across various microclimates [28]. In Vietnam, climate-responsive design and construction strategies ensure necessary climatic comfort [29]. In traditional Tibetan dwellings, optimizing solar radiation and heat storage are passive design strategies for interacting with the climate [30]. Climate-responsive design in rural settlements in China achieves necessary adaptation in the heat of summer and the cold of winter [31]. Studies in India emphasize that passive solar design principles provide thermal comfort in traditional Indian dwellings [32, 33]. Similar studies in Cyprus [34], Italy [35], Greece [36], and Australia [37] show similar results in adhering to climate-responsive passive design principles. A commonality in these studies is the presence of principles and strategies that employ human experience and knowledge in constructing the built environment, adapting to the surrounding climate with minimal or no use of fossil energy.

In Iran, 41% of energy demand is related to the building sector. Heating and cooling are the main sources of energy consumption in the building sector, accounting for 61% of the total energy consumed. Bushehr, a southern port of Iran, has a hot and humid climate, resulting in high electricity consumption due to air-conditioning systems used throughout the year. Despite this, the historical fabric of the city, with a long history of habitation, showcases vernacular architectural principles that create a harmonious relationship between people and their environment. This research aims to study and reinterpret the principles of traditional and vernacular forms in Bushehr that enhance the heat resilience of its residents. After reviewing and summarizing theoretical knowledge on this subject, the principles affecting heat resilience in Bushehr's urban fabric at various scales and layers will be analyzed and summarized.

2 Background Knowledge: Heat Resilience Strategies in Hot and Humid Climates

Air temperature, solar radiation, and wind speed generally have the most significant impact on thermal sensation [6, 38]. In hot and humid areas, the pattern and speed of airflow are crucial for thermal comfort in outdoor environments and directly affect citizens' satisfaction with their presence in urban open spaces. Wind speeds higher than 1 m per second can effectively cool air temperatures above 33.5 °C in hot and humid areas [7]. Accordingly, key strategies for enhancing thermal comfort in urban planning for hot and humid climates include appropriate building layouts, orientation with respect to sunlight, vertical density planning to maximize the benefits of desirable winds, attention to shading, placement of sunshades, design details, building volume, and selection of appropriate construction materials to reduce solar radiation [18, 38].

Urban layout is a key factor in determining the amount of solar radiation received at the ground level and on the facades of urban buildings [39]. Research shows that the dispersion and placement of buildings influence the formation of urban heat islands and enhance urban thermal comfort [40]. Homogeneity of building heights has a desirable effect on ventilation in low-density areas but an undesirable effect in high-density areas [8]. Studies also indicate that the pattern and design of street networks facilitate natural ventilation in hot and humid climates. A study in the UAE found that wind speed depends on the width of the street and the geometry of the buildings. In wider streets, wind speed decreases [41]. The airflow speed at the sidewalk level is slower than at higher street elevations. Therefore, urban setback regulations can enhance ventilation efficiency and thermal comfort in urban street networks [42].

Shading strategies are also crucial. Studies show that north–south streets with medium to high height-to-width ratios provide better comfort conditions in summer, especially in the afternoon, while east–west streets need shading strategies to provide comfort along their edges [43]. A study in Singapore emphasized the importance of shading in improving thermal comfort in hot and humid climates, recommending a height-to-width ratio of 3:1 for optimal shading and thermal comfort in urban streets [44]. In open spaces, the average distance to open areas, type of ground covering, orientation, and complexity are the most significant factors affecting solar radiation absorption [39]. Building geometry and roof type also affect wind flow in street canyons [45]. Houses with covered terraces and higher elevations experience higher wind speeds compared to those with only windows and lower elevations [46].

Materials with high albedo reduce the city's surface temperature, while shade from urban structures and greenery significantly affects the urban microclimate and thermal comfort on sidewalks [40, 47]. Planting trees is the most important and effective technique for reducing heat stress in hot and humid climates, improving thermal comfort by 40–80% [48].

3 Methodology and Case Study

3.1 Methods

This research employs a qualitative approach and logical reasoning to examine passive design strategies for heat resiliency in the historical urban fabric of Bushehr. There are two types of research data: First, library data collected by examining research documents on heat resilience strategies in vernacular and traditional buildings in different cities with hot and humid climates, and specifically, studies conducted on the historical urban fabric of Bushehr and climate adaptation strategies in this context. These data have been content analyzed from more than 40 research articles, and the heat resilience strategies and form solutions considered have been collected and summarized.

The second category of data comprises new and old documents, maps, and images of the historical urban fabric of Bushehr, which have been examined at different scales. Through qualitative analysis, strategies based on the research purpose have been extracted and categorized from these data. The data obtained from these two sources have been qualitatively analyzed to develop a theoretical framework summarizing effective strategies, components, and solutions for heat resiliency in the historical urban fabric of Bushehr.

3.2 Bushehr City as the Case Study

Located in southwest Iran at 28.96° north latitude and 52.83° east longitude, Bushehr is the center of Bushehr province (Fig. 1). It is distinguished as a city with unique vernacular architecture. The traditional urban fabric of Bushehr has grown and expanded in the northern part of Bushehr, which is the most important part of the region from a geographical and spatial point of view (Fig. 1). In this part, the city is surrounded by the sea on three sides—north, east, and west—while the south side is connected to the other parts of the old and new fabric.

Urban development in the Bushehr region dates back to the early civilization of Rishahr, located 12 km from the current city. The history of Rishahr spans from the Elamian period before Islam to the Afsharian period after Islam. During this time, the ports of Rishahr and Liyan were active. As the importance of Rishahr gradually decreased, Liyan village became the location of the new city of Bushehr. The overall structure of old Bushehr was completed during the Qajar period. During the Qajar period, Bushehr Port was formed from four neighborhoods: Shanbadi, Behbahani, Koti, and Dehdashti, with a population of 18,000 in 1867 [49].

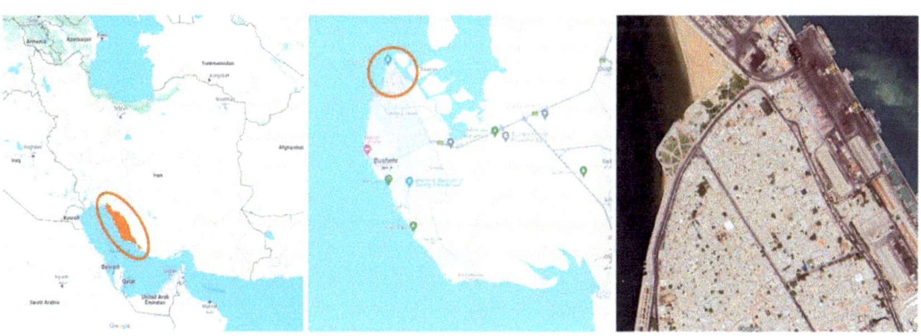

Fig. 1 Location of Bushehr city in Bushehr province and Iran

3.3 Bushehr Climate

With very hot and humid summers and temperate winters, Bushehr is classified as a BWh climate according to Köppen's climate classification. With scarce rainfall, the average minimum and maximum temperatures are 38.1 °C and 10.1 °C, respectively. May, June, July, August, and September are the hottest months in a typical year. For more than 9 months, from April to November, the dry bulb temperature is well above the comfort zone. Since the outdoor dry bulb temperature drops to 8 °C, it is necessary to use thermal mass for the 3 months of December, January, and February (Fig. 2). The average temperature in March, April, and November is within the comfort zone, while the other 6 months of the year are hot.

According to the wind rose chart of this city and the influence of the sea breeze, winds come from different directions, with the prevailing winds being from the north, northwest, and west. With speeds of over 8 m/s, the wind temperatures range from 22 to 28 °C with 30–70% humidity (Fig. 3). Considering the average monthly relative humidity of more than 56%, increasing cross ventilation through shaded windows allows the sea breeze to moderate high relative humidity.

According to the Bushehr Bioclimatic Chart (Fig. 4), only 16.7% of the time are in the comfort zone. However, using adaptive comfort ventilation, 22.9% more of the time would be in the comfort conditions. The best cooling strategies to provide thermal comfort would be:

- Adaptive comfort ventilation (22.9%)
- Sun shading of windows (24.9%)
- Cooling and Dehumidification (39.2%)
- Direct evaporative cooling (2.6%)
- Two-stage evaporative cooling (3.4%)
- Fan-forced ventilation cooling (1.5%)
- High thermal mass (1.4%)
- High thermal mass night flushed (1.7%)

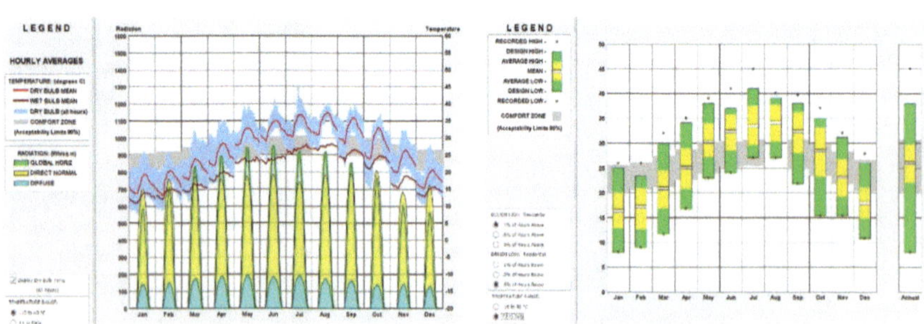

a) Air temperatures and solar radiation b) monthly average dry bulb temperatures

Fig. 2 Climate data of Bushehr (extracted from climate consultant software)

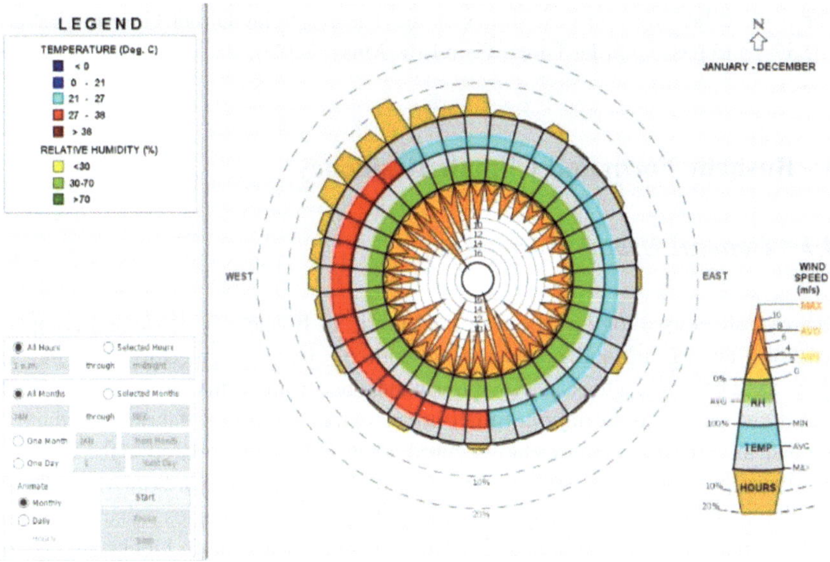

Fig. 3 The wind rose of Bushehr

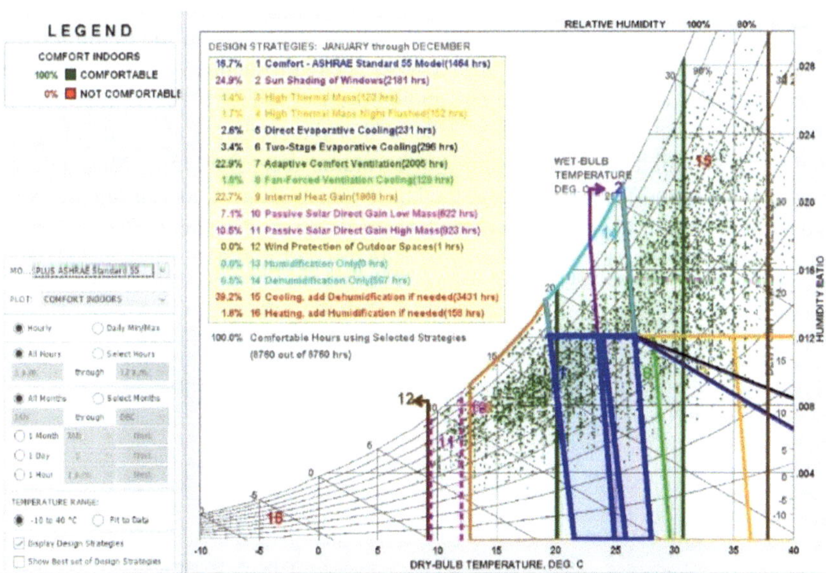

Fig. 4 Bushehr bioclimatic chart (extracted from climate consultant software)

For passive heating, the best strategies are Passive Solar Direct Gain High Mass (10.5%) and passive Solar Direct Gain Low Mass (7.1%).

4 Bushehr Form and Climate Resiliency

4.1 General Urban Morphology

The overall morphology of the historical fabric of Bushehr is dense and compact, featuring an organic and connected network of streets oriented towards the sea (Fig. 5). The compactness facilitates shading and transforms streets into tunnels that draw wind into the fabric. Small urban spaces within the fabric contribute to the distribution of air. Houses have central courtyards but compared to the common courtyard pattern in the hot and dry areas of Iran, the courtyards in Bushehr are smaller, and the surrounding spaces are less deep. These strategies enhance shading and ventilation for the buildings [11, 50]. The small-scale fabric blocks are permeable, minimizing shared walls, which facilitates natural ventilation [51]. The entire fabric functions as a coordinated system, with all components facilitating wind flow and ventilation.

Regarding the street network and its climatic considerations, the first important point is the general orientation of the network. The main openings of the streets align with the prevailing wind from the city and breeze from the sea (Fig. 6) [49, 51, 52]. Simulation and quantitative studies indicate that the entrances and exits of streets and squares exhibit the highest speed and volume of airflow [50]. Masoumi and colleagues selected four streets from the fabric for assessment, measuring the extent and manner of natural ventilation in the historical fabric of Bushehr. The assessment revealed that the orientation of the street network and the proportions of the street

Fig. 5 The compact historical fabric of Bushehr

create air channels at the entrances of the western streets, playing a key role in natural ventilation [11].

A prominent feature of the street network in the historical fabric of Bushehr is its high enclosure and height-to-width ratio (Fig. 7). Studies emphasize the impact of this aspect on increasing the climatic comfort of the fabric. This ratio is on average 5:1 and in some streets reaches up to 10:1, which, in addition to providing appropriate shading, facilitates ventilation performance. The streets have considerable depth and length [49, 50, 51]. The high permeability and connectivity of the network also facilitate airflow. Dead-end streets are rare in the fabric.

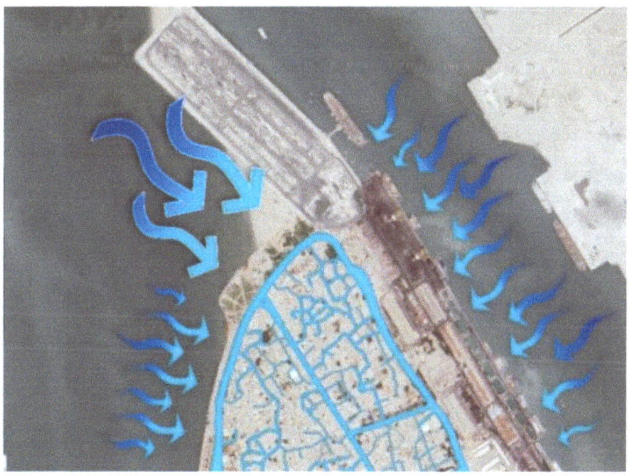

Fig. 6 Wind direction and its flow in the historical fabric of Bushehr

Fig. 7 Enclosure and width-to-height ratio in passages

Fig. 8 Protrusions in passage wall

According to Masoumi et al., dead ends should be minimized to allow free airflow. Sharp angles connecting two paths negatively affect the fluid movement of wind [11]. The street width increases in a funnel shape, which according to the Venturi law can increase wind speed [50]. The street network design features numerous protrusions in the street walls (Fig. 8). These protrusions, sometimes in the form of attachments connected to the facade, roof protrusions, or latticed terraces, are positioned towards the street wall. In some cases, the height hierarchy of the buildings is such that the walls gradually approach each other on the first and second floors [49]. This strategy transforms the street into a wind tunnel, directing airflow to the street level where pedestrians are present. Simulation research by Masoumi and colleagues concluded that changes in path widths and attention to geometry and proportions effectively increase wind speed in the fabric. Adding protruding walls and vegetation in appropriate places has been effective in directing the wind into the depth of the fabric [11].

4.2 Urban Spaces

In the historical fabric of Bushehr, streets guide wind into the fabric, while open spaces strengthen and distribute wind in streets far from the sea [49]. Except for the central spaces of each of the four neighborhoods, other spaces are small-scale, located 20–60 m apart, with distances increasing away from the sea. The small area of space, while providing shade, transforms spaces into a connected network for distributing airflow. Each space has 3 or 4 branches to the fabric, with angles that do not reduce airflow circulation [50]. Research by Masoumi and colleagues, simulating streets and spaces to understand ventilation performance, showed that the location and geometry of spaces facilitate natural ventilation. The orientation of the space considers shading, and the placement of furniture and attachments, as well as vegetation, facilitates airflow [11].

4.3 Spatial Organization of Buildings

Buildings with dense volumes reduce heat loss and provide more shade. The main enclosed spaces with small depths extend around the courtyard with porous walls full of windows and various semi-open spaces [53]. Unlike most parts of Iran, Bushehr's architecture is not completely introverted. Rooms are located around a central courtyard, making buildings semi-introverted with porous facades to link the sea breeze and wind flow from outside to the interior courtyard [54]. The ground floor is designated for service spaces and water storage, while upper-level rooms are primarily used as major spaces. Upper floors are designed to be cooler, with occupants typically residing on these floors to benefit from sea breeze ventilation [55].

4.3.1 Central Courtyard

In Bushehr's historical context, most forms are cube-like porous volumes with expansive porches and projected balconies. Building arrangements include a central courtyard, mainly with an east–west orientation to minimize sun exposure [49]. However, with an average of 15% of the land area, the courtyard is smaller than in central Iran, providing shade during most hours of the day [2, 56, 57]. A typical central courtyard in Bushehr is formed by building on two to four sides [55]. Most houses have two stories with a four-sided square courtyard featuring a high height-to-length ratio. A comparative study of central courtyards in Bushehr and Yazd shows major differences in hot-humid and hot-dry climates [55]. The proportions of the courtyard in Bushehr provide higher shade either on the floor or surrounding facades. Mozafari and Alimardani [58] found that the height ratio in Bushehr courtyards is 2.5 times higher than in hot-dry climates, resulting in more shade during the day. At night, it acts as a heat sink, removing heat quickly. Openings on both external and internal facades assist cross-ventilation, directing wind through room spaces and exiting through the courtyard [58] (Fig. 9).

4.3.2 Semi-open Spaces

Bushehr's architecture uniquely combines open, semi-open, and closed spaces. Semi-open spaces act as buffers between the courtyard and closed spaces, providing shade and ventilation [2]. The combination of space types follows local wind patterns,

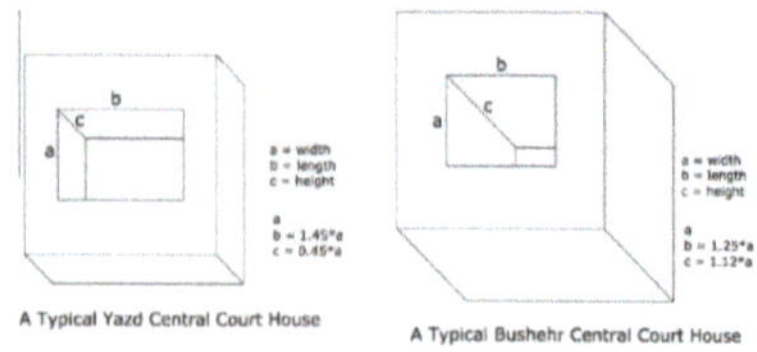

(adapted from: Mozafari and Alimardani, 2020)

Fig. 9 Central courtyards in Bushehr

temperature, and radiation [54]. They act as shading elements to block excessive solar radiation. Semi-open spaces like Tarmeh and Shanashir connect indoor environments to alleys, acting as chimneys to bring out heat through suction [51].

4.3.3 Tarmeh

Tarmeh is a roof-covered space with one side open, used as a seasonal temporary residence or corridor connecting several spaces. It interfaces between open and closed spaces, facilitating natural ventilation [49, 51]. Tarmeh significantly impacts thermal comfort and air quality in adjacent closed spaces [59]. Its climatic effects include reduced received radiation and increased ventilation. Amoozegar et al. [60] defined two types of Tarmeh—internal and external—based on sea orientation to receive prevailing wind (Fig. 10). With lower humidity and higher airflow, Tarmeh is favorable for rest, gathering, daily activities, and temporary seasonal living. It also provides access to other spaces and visual privacy.

a) Exterior Tarmeh b) Intermediate Tarmeh c) Interior Tarmeh d) Uncovered interior Tarmeh

Fig. 10 Different types of Tarmeh

4.3.4 Shanashir

Shanashir, or Shanashil, is a projected wooden balcony on Bushehr's building facades. Using louvered screens, Shanashir blocks sunlight and provides privacy, acts as a light filter, and prevents rapid heat exchange [51, 61]. It facilitates cross-ventilation, provides views, and enhances privacy, similar to Mushrabiyah in Arab architecture. Some jars in Shanashir cool passing air and provide cool water [62]. Different types of Shanashir include exterior, interior, continuous, and non-continuous [62]. Shanashir significantly decreases indoor temperature, reduces energy consumption by 3–4%, and improves daylighting and glare reduction [62, 63]. Openings and wooden shutters allow breeze passage while blocking direct sunlight [2, 49]. Bahrani et al. [61] found that wind vanes increase wind speed and change direction towards upper-floor rooms.

4.3.5 Rooftop

In Bushehr, flat roofs, or "Boon," are used for various activities, including night gatherings and sleep. "Pishboon" refers to similar spaces at lower heights for daily activities. Wooden railing parapets allow sea breeze passage, providing better thermal comfort [51, 58] (Figs. 11, 12).

a) Exterior Shanashir b) Interior Shanashir

Fig. 11 Different types of Shanashir

a) Rooftop and adjacent spaces b) railing parapets

Fig. 12 Views of Rooftop space

4.3.6 Semi-open Staircases

Located in the corners of Bushehr houses, semi-open staircases act as vertical ventilators connecting different levels. They function as solar chimneys and wind outlets, removing hot air from the central courtyard using stack ventilation [51, 56, 5758] (Fig. 13).

a) Connection to other spaces b) Connection to the upper floors c) Connection to the courtyard

Fig. 13 Views of semi-open Staircase

4.3.7 Openings

In the central part of Iran, there are no windows in the exterior facades. But in Bushehr, a large number of openings are placed on the external and internal facades, to increase wind flow speed [49, 58]. Having openings on two opposite sides of a room assists in natural cross-ventilation [55]. Considering their important role in directing wind and natural ventilation, openings are formed at different levels with diverse forms. They are placed in the wind direction to capture the sea breeze. Houses have several openings both on the interior and exterior facades to facilitate natural ventilation. Wind flow is facilitated through many openings and vertical connections between openings at different levels [58]. Openings to public spaces are usually covered by wooden blinds and shutters to control views from outside while allowing wind into the rooms (Fig. 14) [5, 55] (Fig. 15).

Fig. 14 Opening types and arrangements in Bushehr vernacular houses

Fig. 15 The light colors of the facades and colorful windows

4.3.8 Shadings

Various strategies minimize sunlight penetration through openings. Shading devices are installed on balconies and windows, especially facing south. Wooden shutters on windows control solar overheating and ventilation [49]. Openings are usually placed in thick walls with wooden louvers to block sun radiation, adjustable to control wind direction and amount, reducing indoor humidity.

4.3.9 Materials and Colours

In hot and humid regions, low-heat capacity materials are most suitable. In Bushehr, wood, straw, limestone, coral stone, and sedimentary rock are used for their low heat capacity and accessibility [51]. Porous materials provide ventilation and rapid moisture drying from exterior walls [2, 50]. Porous limestone, or "shemsh," is extensively used. Sunshades, doors, and windows are made of wood, straw, and stained glass. "Chandal" beams, brought from India or Africa, cover ceilings in rooms, corridors, and porches [53]. Openings often feature "Orosi," colorful courtyard-facing doors and windows that open upwards, made of wood and stained glass to reduce heat radiation. Building surfaces are light-colored to reflect sunlight, with most facades colored white for radiative cooling and light reflection. Coral stone covered with white plaster is commonly used [64].

5 Discussion

Heat resilience in the historical fabric of Bushehr can be conceptualized as a system operating across three scales—micro, meso, and macro—with numerous components and solutions aimed at fulfilling three main strategies: thermal control, active and natural ventilation, and shading (Table 1 and Fig. 16). As seen in Table 1, at the micro, meso, and macro scales, these three main strategies are critical for heat resilience in traditional and vernacular construction. At the macro scale, the general morphology of the historical fabric, with its density, specific orientation, and vertical layout, enhances heat resiliency. The fabric's density and low ratio of open to enclosed spaces provide street shading. The general orientation takes advantage of wind flow and minimizes solar radiation effects throughout the day. Additionally, varied building heights increase wind speed within the fabric.

At the meso scale, the street and block network components contribute to the fabric's heat resilience. Connected streets act like channels directing prevailing winds and sea breezes deep into the fabric. The east–west street orientation focuses on shading and thermal control. The number and angles of intersections ensure adequate ventilation. The high ratio of height to width facilitates shading and ventilation. Occasional street protrusions intensify airflow. Small-scale blocks enclosed by narrow streets facilitate deep ventilation. At the micro scale, central courtyards with a low

Table 1 Components, solutions and Strategies in Heat Resiliency System

Scale	Component	Policies/ solutions	Description	Heat resilience strategy		
				Heat avoidance/ heat control	Natural/ passive ventilation	Shading
Macro	Urban fabric	Compactness	Low ratio of open to closed space to increase shading	✓	–	✓
		Orientation	Considering the direction of the wind and sunlight		✓	✓
		Height density	The variation in height increases the wind speed	✓	✓	–
Messo	Urban network	Connectivity	Facilitate ventilation	–	✓	–
		Orientation	Considering the direction of the wind and sunlight	✓	✓	✓
		Width-to-height ratio	High closures: facilitate shading	–	✓	✓
		Intersections	Consideration of ventilation in the number and angles of intersections	–	✓	–
	Blocks	Protrusions in the façade wall	In order to Increasing wind speed	–	✓	–
		Small blocks	Permeability and facilitation of ventilation and shading	–	✓	–
		Block isolating	Placing passageways around the blocks and facilitating ventilation	–	✓	–
Micro	Buildings	Courtyard	Depth-to-length low ratio of spaces around the yard	✓	–	–

(continued)

Table 1 (continued)

Scale	Component	Policies/ solutions	Description	Heat resilience strategy		
				Heat avoidance/ heat control	Natural/ passive ventilation	Shading
		Semi introvert	Openings in exterior and interior facades	✓	–	–
		Floors	Number of stories-service spaces on the ground floor	✓	–	–
	Semi-open spaces	Tarmeh	A roof-covered space with one side open	✓	✓	✓
		Shanashir	A wooden balcony in Bushehr's building facades	✓	✓	✓
		Boon	Flat roofs that used for various activities		✓	
		Staircase	Semi open staircases are used as vertical ventilators to connect different levels	✓	✓	✓
	Openings	Windows	Openings on both internal and external facades, High WWR, Vertical windows	–	✓	✓
	Shadings	Straw sheets, louvers		–	✓	✓
	Materials	Low heat capacity	Coral stone, mud, gypsum, limestone, Chandal wood, Straw sheets, Tesk stone	✓	✓	–

(continued)

Table 1 (continued)

Scale	Component	Policies/ solutions	Description	Heat resilience strategy		
				Heat avoidance/ heat control	Natural/ passive ventilation	Shading
		Light color materials	To reflect great part of solar radiation	✓	–	✓
		Stained glass	To absorb small part of solar radiation	✓		✓
	Vegetation	In courtyard	Few trees sometimes with small fountain	–	✓	–

open-to-enclosed space ratio aid shading and ventilation. Greenery in the courtyard contributes to thermal comfort. The semi-introverted system, with openings to both courtyard and street, aids natural ventilation. Semi-open spaces like roofed terraces, jalousies, and external staircases enhance natural ventilation and thermal control. Window ratio, dimensions, and louvre shading devices optimize ventilation and shading. Light-colored materials reduce solar radiation and heat absorption. The matte finish on glass controls heat and sunlight. Notably, the fabric's heat resilience strategies operate across all three scales and components, forming a coherent system aimed at enhancing heat resilience.

6 Concluding Remarks

This research aims to review vernacular heat resiliency strategies in historical city construction to inform the planning, design, and architecture of new settlements. We reviewed literature, research, documents, images, and maps, selecting the historical urban fabric of Bushehr as a case study. The key heat resiliency strategies and solutions were summarized and presented in this framework.

This framework categorizes the components, solutions, and strategies for building architecture and urban planning and design for heat resiliency in Bushehr into three scales: macro, meso, and micro. Generally, shading, natural ventilation, and heat control are the most important strategies for promoting heat resiliency in this historical area. At the macro scale, we explored the general characteristics of the historical fabric; at the meso scale, we examined the urban network and blocks; and at the micro-scale, we analyzed architectural components, open spaces, greenery, and materials.

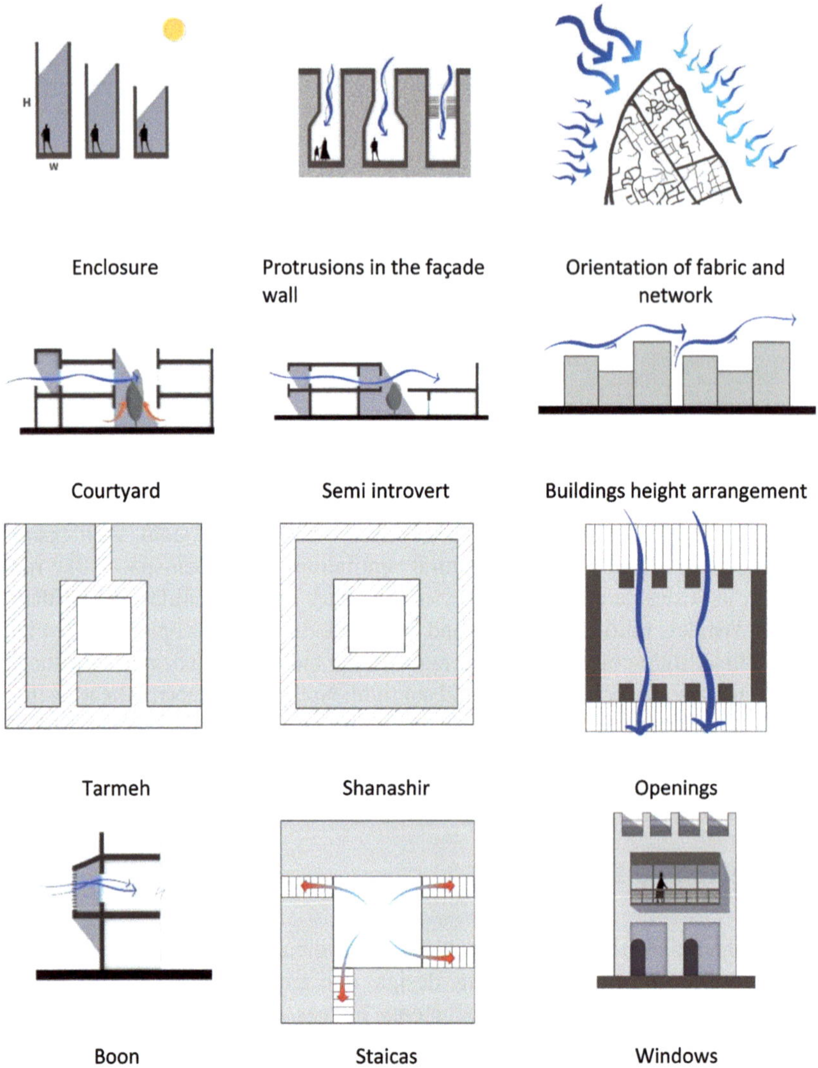

Fig. 16 Some of the heat resiliency solutions in Bushehr's historical fabric

Key solutions for improving heat resiliency include compactness, orientation of the fabric and passage network, enclosure, intersections, block size and permeability, central courtyards, semi-open spaces like Tarameh and Shanashir, design and proportions of openings and shades, the color and thermal capacity of materials, and greenery in courtyards. This research can inform the planning and design of new settlements, especially in hot and humid areas, to adopt heat-resiliency policies in architecture and urban planning.

References

1. Indraganti M (2010) Understanding the climate sensitive architecture of Marikal, a village in Telangana region in Andhra Pradesh, India. Energy Build 45:2709–2722
2. Mohammadi A, Saghafi MR, Tahbaz M, Nasrollahi F (2018) The study of climate-responsive solutions in traditional dwellings of Bushehr City in Southern Iran. J Build Eng 16:169–183
3. Yang W, Xu J, Lu Z, Yan J, Li F (2022) A systematic review of the indoor thermal environment of the vernacular dwelling climate responsiveness. J Build Eng 53:104514
4. Karimi B (2012) Investigating the effects of Bushehr old designing on architecture of Persian Gulf border countries. Hoviat-E-Shahr 6:85–96 (**in Persian**)
5. Shahin AT, Baghaei S (2006) Sustainability patterns in the old residential fabric of Bushehr. Archit Constr 7(10):130–135 (**in Persian**)
6. Villadiego K, Velay-Dabat MA (2014) Outdoor thermal comfort in a hot and humid climate of Colombia: a field study in Barranquilla. Build Environ 75:142–152
7. Yan H, Liu Q, Zhao W, Pang C, Dong M, Zhang H, Gao J, Wang H, Hu B, Yang L, Wang L (2020) The coupled effect of temperature, humidity, and air movement on human thermal response in hot–humid and hot–arid climates in summer in China. Build Environ 177:106898
8. Wang W, Ng E, Yuan C, Raasch S (2017) Large-eddy simulations of ventilation for thermal comfort—a parametric study of generic urban configurations with perpendicular approaching winds. Urban Clim 20:202–227
9. Bouyer J, Inard C, Musy M (2011) Microclimatic coupling as a solution to improve building energy simulation in an urban context. Energy Build 43(7):1549–1559
10. Freire RZ, Abadie MO, Mendes N (2013) On the improvement of natural ventilation models. Energy Build 62:222–229
11. Masoumi HR, Nejati N, Ahadi A (2017) Learning from the heritage architecture: developing natural ventilation in compact urban form in hot-humid climate: case study of Bushehr, Iran. Int J Archit Herit 11(3):415–432
12. Taleb HM (2015) Natural ventilation as energy efficient solution for achieving low-energy houses in Dubai. Energy ings 99:284–291
13. Aldossary NA, Rezgui Y, Kwan A (2017) Establishing domestic low energy consumption reference levels for Saudi Arabia and the Wider Middle Eastern Region. Sustain Cities Soc 28:265–276
14. Taleb HM (2014) Using passive cooling strategies to improve thermal performance and reduce energy consumption of residential buildings in UAE buildings. Front Archit Res 3(2):154–165
15. Oropeza-Perez I, Østergaard PA (2014) Energy saving potential of utilizing natural ventilation under warm conditions—a case study of Mexico. Appl Energy 130:20–32
16. Huang KT, Hwang RL (2016) Future trends of residential building cooling energy and passive adaptation measures to counteract climate change: the case of Taiwan. Appl Energy 184:1230–1240
17. Toe DHC, Kubota T (2015) Comparative assessment of vernacular passive cooling techniques for improving indoor thermal comfort of modern terraced houses in hot–humid climate of Malaysia. Sol Energy 114:229–258
18. Chandel SS, Sharma V, Marwah BM (2016) Review of energy efficient features in vernacular architecture for improving indoor thermal comfort conditions. Renew Sustain Energy Rev 65:459–477
19. Tzikopoulos AF, Karatza MC, Paravantis JA (2005) Modeling energy efficiency of bioclimatic buildings. Energy Build 37(5):529–544
20. Pan Y, Zhong W, Zheng X, Xu H, Zhang T (2024) Natural ventilation in vernacular architecture: a systematic review of bioclimatic ventilation design and its performance evaluation. Build Environ 253:111317
21. Xu H, Huang Q, Liu G, Zhang Q (2016) A quantitative study of the climate-responsive design strategies of ancient timber-frame halls in northern China based on field measurements. Energy Build 133:306–320

22. Engin N, Vural N, Vural S, Sumerkan MR (2007) Climatic effect in the formation of vernacular houses in the Eastern Black Sea region. Build Environ 42(2):960–969
23. Zhai ZJ, Previtali JM (2010) Ancient vernacular architecture: characteristics categorization and energy performance evaluation. Energy Build 42:357–365
24. Borong L, Gang T, Peng W, Ling S, Yingxin Z, Guangkui Z (2004) Study on the thermal performance of the Chinese traditional vernacular dwellings in Summer. Energy Build 36:73–79
25. Dili AS, Naseer MA, Varghese TZ (2010) Passive environment control system of Kerala vernacular residential architecture for a comfortable indoor environment: a qualitative and quantitative analyses. Energy Build 42:917–927
26. Dili AS, Naseer MA, Varghese TZ (2011) Passive control methods for a comfortable indoor environment: comparative investigation of traditional and modern architecture of Kerala in summer. Energy Build 43:653–664
27. Priya RS, Sundarraja MC, Radhakrishnan S, Vijayalakshmi L (2012) Solar passive techniques in the vernacular buildings of coastal regions in Nagapattinam, Tamil Nadu-India—a qualitative and quantitative analysis. Energy Build 49:50–61
28. Bodach S, Lang W, Hamhaber J (2014) Climate responsive building design strategies of vernacular architecture in Nepal. Energy Build 81:227–242
29. Nguyen A, Tran Q, Tran D, Reiter S (2011) An investigation on climate responsive design strategies of vernacular housing in Vietnam. Energy Build 46:2088–2106
30. Huang L, Hamza N, Lan B, Zahi D (2016) Climate-responsive design of traditional dwellings in the cold-arid regions of Tibet and a field investigation of indoor environments in winter. Energy Build 128:697–712
31. Gou S, Li Z, Zhao Q, Nik Q, Scartezzini J (2015) Climate responsive strategies of traditional dwellings located in an ancient village in hot summer and cold winter region of China. Build Environ 86:151–165
32. Priya S, Sundarraja MC, Radhakrishnan S, Vijayalakshmi L (2010) Solar passive techniques in the vernacular buildings of coastal regions in Nagapattinam Tamil Nadu-India—a qualitative and quantitative analysis. Energy Build 49:50–61
33. Singh MK, Mahapatra S, Atreya SK (2011) Solar passive features in vernacular architecture of North-East India. Sol Energy 85:2011–2022
34. Michael A, Demosthenous D, Philokypro M (2017) Natural ventilation for cooling in mediterranean climate: a case study in vernacular architecture of Cyprus. Energy Build 144:333–345
35. Desogus G, Cannas LGF, Sanna A (2016) Bioclimatic lessons from Mediterranean vernacular architecture: the Sardinian case study. Energy Build 129:574–588
36. Oikonomou A, Bougiatioti F (2011) Architectural structure and environmental performance of the traditional buildings in Florina, NW Greece. Energy Build 46:669–689
37. Bre F, Fachinotti VD (2017) A computational multi-objective optimization method to improve energy efficiency and thermal comfort in dwellings. Build Environ 154:283–294
38. Ng E, Cheng L (2012) Outdoor thermal comfort and outdoor activities: a review of research in the past decade. Cities 29(2):118–125
39. Chatzipoulka C, Compagnon R, Nikolopoulou M (2016) Urban geometry and solar availability on façades and ground of real urban forms: using London as a case study. Sol Energy 138:53–66
40. Jamei E, Rajagopalan P, Seyedmahmoudian M, Jame Y (2016) Review on the impact of urban geometry and pedestrian level greening on outdoor thermal comfort. Renew Sustain Energy Rev 54:1002–1017
41. Al-Sallal KA, Al-Rais L (2012) Outdoor airflow analysis and potential for passive cooling in the modern urban context of Dubai. Renew Energy 38(1):40–49
42. da Silva FT, de Alvarez CE (2015) An integrated approach for ventilation's assessment on outdoor thermal comfort. Build Environ 87:59–71
43. Achour-Younsi S, Kharrat F (2016) Outdoor thermal comfort: impact of the geometry of an urban street canyon in a Mediterranean subtropical climate—case study Tunis, Tunisia. Proc Soc Behav Sci 216:689–700

44. Yang W, Wong NH, Lin Y (2015) Thermal comfort in high-rise urban environments in Singapore. Proc Eng 121:2125–2131
45. Llaguno-Munitxa M, Bou-Zeid E, Hultmark M (2017) The influence of building geometry on street canyon air flow: validation of large eddy simulations against wind tunnel experiments. J Wind Eng Ind Aerodyn 165:115–130
46. Bay E, Martinez-Molina A, Dupont W (2022) Assessment of natural ventilation strategies in historical buildings in a hot and humid climate using energy and CFD simulations. Build Eng 51(1):10428
47. Chatzidimitriou A, Yannas S (2015) Microclimate development in open urban spaces: the influence of form and materials. Energy Build 108:156–174
48. Yang SR, Lin TP (2016) An integrated outdoor spaces design procedure to relieve heat stress in hot and humid regions. Build Environ 99:149–160
49. Ranjbar E, Pourjafar MR, Khaliji K (2010) Innovations in climatic designing due to the wind flowing through the old Bushehr. Bagh Nazar 13:17–34 (**in Persian**)
50. Shaeri J, Aflaki A, Yaghoubi M, Janalizadeh H (2018) Investigation of passive design strategies in a traditional urban neighborhood: a case study. Urban Clim 26:31–50
51. Motealleh P, Zolfaghari M, Parsaee M (2016) Investigating climate responsive solutions in vernacular architecture of Bushehr city. Hous Build Natl Res Center HBRC J HBRC J 14(2):215–223
52. Asghari M, Vafaei R (2015) Comparative analysis of climate and morphology between traditional patterns and modern approaches: the case of Bushehr's traditional context. Proc Eng 118:590–605
53. Nadoomi R, Sharghi A, Nakhaei S, Azadian R (2023) Regional materials and environmental sustainability in hot and humid climates: a study on Boushehr's vernacular houses. Int J Archit Eng Urban Plan 33(4):1–19
54. Afsari M, Yousefi R (2020) Investigating the effect of climate on warm and humid architecture, a review of the contextualism approach in native architecture of Bandar Abbas and Bushehr regions. Int J Eng Technol (IJERT) 9(12):31–36
55. Khajehzadeh I, Vale B, Yavari F (2016) A comparison of the traditional use of court houses in two cities. Int J Sustain Built Environ 5:470–483
56. Nikghadam N (2015) Climatic patterns of functional spaces in vernacular houses of bushehr. Bagh e Nazar 12(39):29–42 (**in Persian**)
57. Nikghadam N (2022) Climatic typology of vernacular houses along the northern shores of the Persian gulf and the Oman sea: how to follow paradigms in contemporary housing design. Soffeh 32(3):31–47 (**in Persian**)
58. Mozafari N, Alimardani M (2020) Climate adaptability of old and new house in Bushehr's historical texture. Civ Environ Eng 16(2):249–258
59. Ghahraman Izadi N, Taghipour M, Eskandari H, Movahed K (2023) The role of the effective parameters of Taremeh in improving natural ventilation in the courtyard houses of Bushehr in the historical context. Bagh Nazar 125(20):23–42 (**in Persian**)
60. Amuzegar M, Vakilinezhad R, Aliabadi M (2021) Typology of Tarmeh as a semi-open space in vernacular architecture of Bushehr. Pazhoheshha-ye Bastan Shenasi Iran 11(30):271–290 (**in Persian**)
61. Bahrani HR, Ranjbar E, Ahmadi S (2018) An analysis of the function of external Senasir types and their effects on the wind flow in the historical part of Bushehr. Naqshejahan Basic Stud New Technol Archit Plan 8(3):179–186 (**in Persian**)
62. Mofidi SM, Vakilinezhad R, Mehdizadeh F (2013) Shanashil: a sustainable element to balance light, view, and thermal comfort. Int J Environ Sustain 8(4):101–110
63. Alrashed F, Asif M, Burek S (2017) The role of older construction techniques and materials for developing zero-energy homes in various desert climates. Buildings 7(1):17
64. Rasaie Kashuk S (2005) Glory of abushahr. Tolou Bushehr (**in Persian**)

Adaptive Pathways to Flood Resilient Infrastructure in African Coastal Cities: Lessons and Experiences from Nigeria and South Africa

Olasunkanmi Habeeb Okunola

Abstract The recent surge in flood occurrences in certain cities of the Global South, particularly in sub-Saharan Africa (SSA), highlights the inadequate preparedness and infrastructure of these cities in dealing with the consequences of extreme weather-induced hazards. Like many others worldwide, Sub-Saharan African cities face the dual challenge of rapid urbanisation and deficiencies in governance, public institutions, and infrastructure development, undermining their efforts to build healthy, climate-resilient cities. Consequently, this chapter aims to bridge this gap by examining how current infrastructure governance aligns with addressing present and projected complex, cascading, and systemic risks arising from climatic extreme flood events in Lagos, Nigeria, and Durban, South Africa. Employing a case study approach, a combination of concurrent triangulation mixed methods, including in-depth interviews, survey questionnaires, focus group discussions, and seminars, was utilized to gather data from government agency representatives, civil society organisations, community members, and academia. In addition, text-mining techniques were applied to assess relevant policy documents and frameworks. Results show that Lagos, Nigeria, and eThekwini, South Africa, face significant challenges in aligning their current infrastructure governance with the complex and evolving risks of climatic conditions. Inadequate infrastructure planning, limited coordination among stakeholders, and a lack of implementation of relevant policies and frameworks were identified as key obstacles. The chapter proposes pathways that promote integrated governance in providing and managing flood-resilient infrastructure. These pathways include (i) promoting inclusiveness in the implementation of critical infrastructure, (ii) facilitating knowledge co-creation and adaptive capacity among stakeholders, (iii) meeting long and short-term infrastructure needs, and (iv) accounting for complexity and long-term changes in infrastructure development, including the potential need for transformational change. These themes have been explored within the broader framework of the 2030 Sustainable Development Goals,

O. H. Okunola (✉)
Vulnerability Assessment, Risk Management and Adaptive Planning Section, United Nations University-Institute for Environment and Human Security, Bonn, Germany
e-mail: olasunkanmi.habeeb@gmail.com

A. Cheshmehzangi et al. (eds.), *Designing Healthy Cities*, Urban Sustainability,
https://doi.org/10.1007/978-981-96-5538-0_7

underscoring the critical importance of building healthy, climate-resilient cities in sub-Saharan Africa.

Keywords Critical infrastructure · Flood risk · Cascading risk · City resilience · Adaptive pathways

1 Introduction

Climate-related extreme events pose significant threats to vulnerable communities, ecosystems, livelihoods, and critical infrastructure around the world [1]. Among these events, floods remain one of the most devastating in some cities of the Global South, particularly in sub-Saharan Africa (SSA) [2]. They have severe direct and indirect impacts on health, social, economic, and financial systems [3, 4]. These challenges are further compounded by rapid urbanization in the context of economic stagnation, poor governance, fragile public institutions, and inadequate infrastructural development [5]. The World Economic Forum [6] has observed that Africa's population is projected to double by 2050, with more than 80% of the increase occurring in cities, especially slums. This unprecedented growth has brought about various problems, including unemployment, the informal sector's growth, deteriorating critical infrastructure, overcrowding, environmental degradation, and a housing shortage [7].

In line with these urban socio-economic challenges [8] argue that the absence of political will has resulted in the unplanned spatial expansion of many cities, leaving city managers and planners overwhelmed. Consequently, cities and their inhabitants, as well as their infrastructure, have become highly vulnerable to the frequency of flood episodes. This chapter focuses on the significant impacts of frequent extreme flood events on infrastructure development in cities across sub-Saharan Africa. This infrastructure includes flood barriers and walls, flood gates and doors, raised foundations, green infrastructure, floodwater drainage systems, and flood-resistant materials. These infrastructure systems have experienced significant damage, leading to disruptions in critical services, increased vulnerability of communities, and economic losses [2].

For instance, it is estimated that almost US$200 million per year of infrastructure in sub-Saharan Africa, including coastal and rainwater infrastructure, is at risk from flooding, with Angola, Ghana, Nigeria, and South Africa presenting the highest risk [2]. The lack of adequate flood-resilient infrastructure exacerbates the negative consequences of these flood events, underscoring the urgent need for investment in robust and adaptive infrastructure solutions to enhance the resilience of cities in sub-Saharan Africa to future flood risks [9]. Consequently, this development requires a sequence of developmental actions to promote the adaptability of flood infrastructure in flood-prone communities, enabling them to accommodate changing conditions over time.

In this regard, studies such as [10, 11] have posited adaptive pathways to represent requisite actions that must be implemented to address uncertainty related to extreme events and prevent maladaptive consequences. Pathways in this context refer to sequences of actions by multiple stakeholders in infrastructure governance to prevent, mitigate, and manage frequent flooding. Adaptive pathways for resilient infrastructure seek innovative approaches to address knowledge gaps and highlight adaptive pathway solutions that foster resilience and sustainability of infrastructure systems in changing environments [12, 13]. Implementing adaptive pathways for resilient infrastructure requires adopting short-term strategies involving multiple stakeholders and progressive implementation for critical infrastructure such as transportation networks, water supply systems, and energy grids. This approach integrates policies, actors, levels, and sectors, inspiring changes in societal debates, institutional settings, and participatory approaches [14]. Effective adaptive pathways for flood-resilient infrastructure entail institutional processes that function across sectors and promote social learning.

While several research studies have examined various adaptive pathways in managing and responding to address disasters and climate change in cities of the Global South and North [15, 16, 17], there is limited understanding of how these pathways can be effectively applied to address the unique challenges faced by cities vulnerable to the impacts of climate change, especially in sub-Saharan Africa regarding infrastructure development. Therefore, the primary objective of this chapter is to address this knowledge gap by exploring and evaluating the potential adaptive pathways for flood-resilient infrastructure in Lagos, Nigeria, and Durban, South Africa.

The research questions that guide this study are as follows: What are the existing challenges and vulnerabilities in the flood-prone cities of Lagos, Nigeria, and Durban, South Africa? How have other Global South and North cities implemented adaptive pathways for flood-resilient infrastructure, and what are the key lessons learned? Considering their specific contexts, what are the potential adaptive pathways for flood-resilient infrastructure in Lagos, Nigeria, and Durban, South Africa? How can the identified adaptive pathways be integrated into policy and planning processes to enhance resilience and promote sustainable development in these cities?

By addressing these research questions, this chapter aims to provide valuable insights and recommendations for policymakers, urban planners, and stakeholders in sub-Saharan African cities to develop effective adaptive pathways for flood-resilient infrastructure.

The chapter is structured as follows: it provides an overview of the study area, describes the methodology used, presents the key results, discusses the potential adaptive pathways for the two cities, and concludes with recommendations on how the identified pathways can be integrated into policy and planning processes.

1.1 Study Areas

This study focuses on two African coastal cities, Lagos and Durban, selected for their history of annual flooding (Fig. 1a, b).

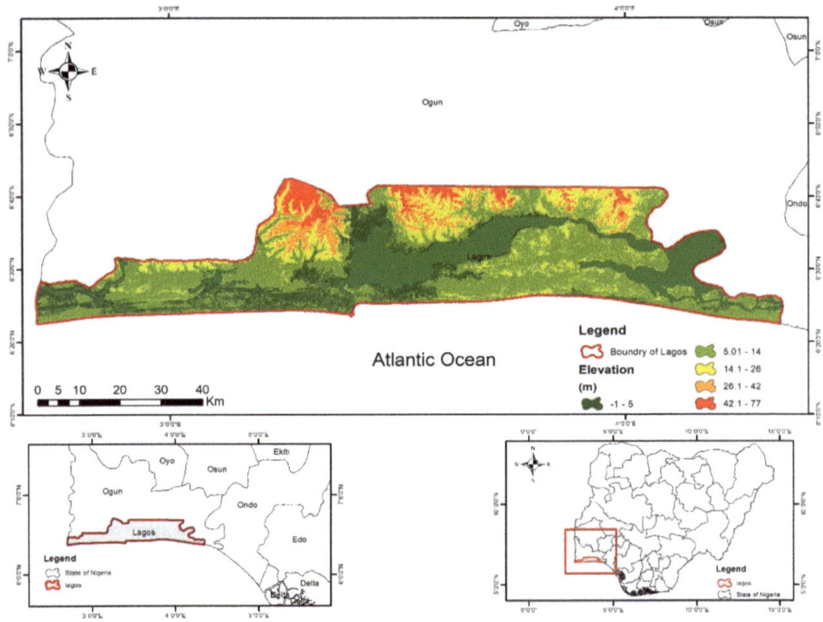

Geographic location of Lagos (Source: Produced by the author)

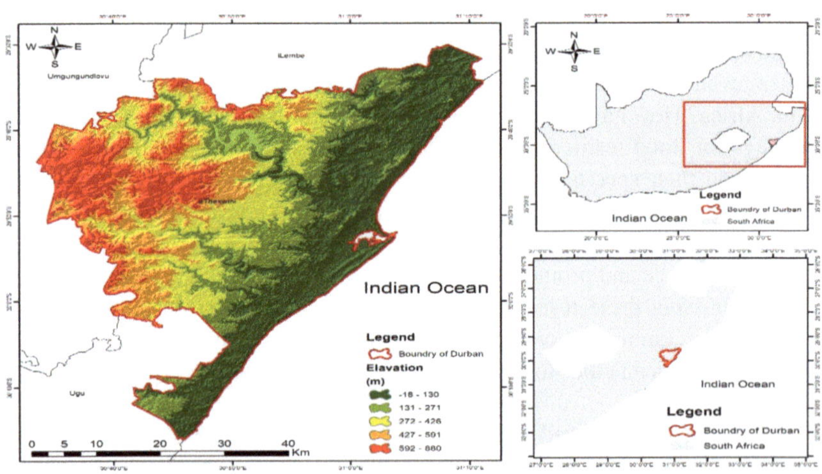

Geographic location of Durban (Source: Produced by the author)

Fig. 1 **a** Geographic location of Lagos. *Source* Produced by the author. **b** Geographic location of Durban. *Source* Produced by the author

Lagos, located in southwestern Nigeria along the Atlantic coastline, is a rapidly growing and densely populated city that faces significant infrastructure challenges, particularly when it comes to flooding [18]. The city's existing flood infrastructure includes drainage systems, flood control channels, embankments, and floodgates. However, rapid urbanization, poor planning, and inadequate maintenance often overwhelm and damage these systems during flood events. Lagos' low-lying topography and heavy rainfall, coupled with tidal surges, further exacerbate its vulnerability. Informal settlements and low-income neighborhoods are at high risk, lacking access to adequate flood resilience infrastructure and resources. While Lagos has recognized the need for improved flood-resilient infrastructure and has undertaken various adaptation efforts, such as constructing new drainage systems, rehabilitating existing infrastructure, and exploring green infrastructure approaches, the city continues to experience devastating floods that cause significant harm to people, critical infrastructure, and have cascading effects.

Durban, situated on the eastern coast of South Africa, is a major port city with a well-established network of roads, bridges, buildings, and drainage infrastructure. However, like Lagos, Durban faces significant challenges in dealing with the impacts of flooding, particularly during intense rainfall events [19]. The city is vulnerable to urban and riverine flooding, with low-lying areas and informal settlements at higher risk [20]. Factors contributing to Durban's flood vulnerability include heavy rainfall, inadequate drainage systems in certain areas, and the effects of climate change, such as sea-level rise [21]. Durban has implemented various adaptation measures to enhance its flood resilience, including improving drainage infrastructure, constructing flood mitigation structures, and developing early warning systems. The city has also explored nature-based solutions, such as wetland restoration and green infrastructure, to manage flood risks and increase its resilience. Additionally, Durban has implemented its climate strategy to strengthen its overall resilience against climate impacts. Despite these efforts, extreme floods continue to impact the city's infrastructure and the livelihoods of its residents every year. This suggests that while progress has been made, there are ongoing challenges and a need for continuous adaptation and improvement to address the persistent flood risks faced by Durban effectively.

2 Methodology

This study adopts a case study approach to provide a contextualized and in-depth understanding of how flood-resilient infrastructure has been implemented and managed to address flooding in Lagos, Nigeria, and Durban, South Africa. The case study approach allows for a detailed examination of the drivers, governance structures, and strategies to prevent and mitigate flooding by implementing resilient infrastructure. The data for this study were collected in Lagos and Durban between November 2022 and May 2023. The researcher interviewed representatives from community groups, government agencies, NGOs, and the private sector to gather insights into the challenges, adaptations, and ongoing efforts to enhance flood resilience in these two coastal cities.

2.1 Data Collection

The data collection process for this study employed a concurrent triangulation mixed methods research design to gather a comprehensive dataset on the implementation of resilient infrastructure for flood risk management in Lagos and Durban. In the initial stage, a thorough review of existing policy documents and frameworks was conducted to understand the governance and infrastructure mechanisms in place for resilient infrastructure in the two case study cities. This review examined policies such as the Lagos State Resilience Strategy and Durban Climate Strategy, providing a solid research foundation.

The mixed methods approach included the following components:

Survey Questionnaires
A total of 270 questionnaires were administered across four coastal communities prone to flooding in each city—Lagos (Ikorodu, Ajah, Victoria Island, Badagry) and Durban (Umhlanga, Bluff, Amanzimtoti, Ballito). These communities were selected based on their differing socioeconomic compositions and geographic locations within the cities. The author and 17 trained enumerators administered the questionnaires, who targeted randomly selected household heads to ensure representative coverage.
In-Depth Interviews (IDIs)
In-depth interviews were conducted with 32 participants in Lagos and 26 in Durban. The interviewees included residents as well as representatives from government agencies and non-governmental organizations (NGOs), identified during the survey process and recommended by some survey respondents.
Focus Group Discussions (FGDs)
Eighteen people participated in focus group discussions in Lagos, and 12 participated in Durban. The participants for these discussions were drawn from the same stakeholder groups as the in-depth interviews.
Workshops
Workshops were organized to facilitate collaborative dialogues on the implementation of flood-resilient infrastructure. These workshops included policymakers, NGOs, private companies, and research organizations.

The collected data aimed to capture information on the following aspects:

- Cascading impacts of flood events in the case studies
- Condition and capacity of various infrastructures to withstand frequent flooding.
- Policy frameworks and guidelines related to flood-resilient infrastructure
- Barriers and challenges to the implementation of flood-resilient infrastructure.
- Community engagement and public participation in flood resilience efforts.

2.2 Data Analysis

Data analysis encompassed both qualitative and quantitative methods. Text mining techniques were applied to assess relevant policy documents and frameworks, providing insights into the policy landscape and supporting the findings derived from the primary data sources. Thematic analysis was conducted on qualitative data, including interview transcripts and focus group discussions, to identify key themes, patterns, and emerging insights. This analysis resulted in the development of hierarchically organized key categories and subcategories through an iterative and inductive process.

Quantitative analysis was performed on the data obtained from survey questionnaires. Descriptive analysis techniques were utilized to derive meaningful statistics from the quantitative data. The combined analysis of qualitative and quantitative data provided a comprehensive understanding of the challenges, strategies, policies, and community involvement related to resilient infrastructure in Lagos, Nigeria, and Durban, South Africa. The findings contribute to the knowledge base on effective flood resilience practices and inform recommendations for policy and practice in coastal cities facing similar challenges.

3 Results

3.1 Cascading Impacts of Flood Events in the Case Studies

Findings from Fig. 2 of the case studies indicate that respondents' views on the cascading impacts of floods in Lagos and Durban align with the frequency-based analysis. In Lagos, the most prevalent cascading impacts identified by respondents are infrastructure damage (28.99%), displacement of residents (19.72%), and economic losses (19.48%). These findings highlight the susceptibility of Lagos' infrastructure to flood events and the resulting social and economic disruptions, emphasizing the urgent need for enhanced resilience measures to create healthier and more climate-resilient communities.

Similarly, in Durban, respondents identified infrastructure damage (23.86%), economic losses (20.05%), and environmental impacts (17.26%) as the primary consequences of flooding. These findings reaffirm the vulnerability of Durban's infrastructure to floods and the associated economic repercussions, which can undermine efforts to develop healthy and climate-resilient buildings and neighborhoods. The significant proportion of environmental impacts underscores the importance of addressing coastal erosion and habitat destruction in the city, a crucial aspect of building climate-resilient communities.

These findings shed light on the shared challenges faced by Lagos and Durban in managing the impacts of flooding. Respondents' recognition of infrastructure damage, economic losses, and environmental impacts underscores the need

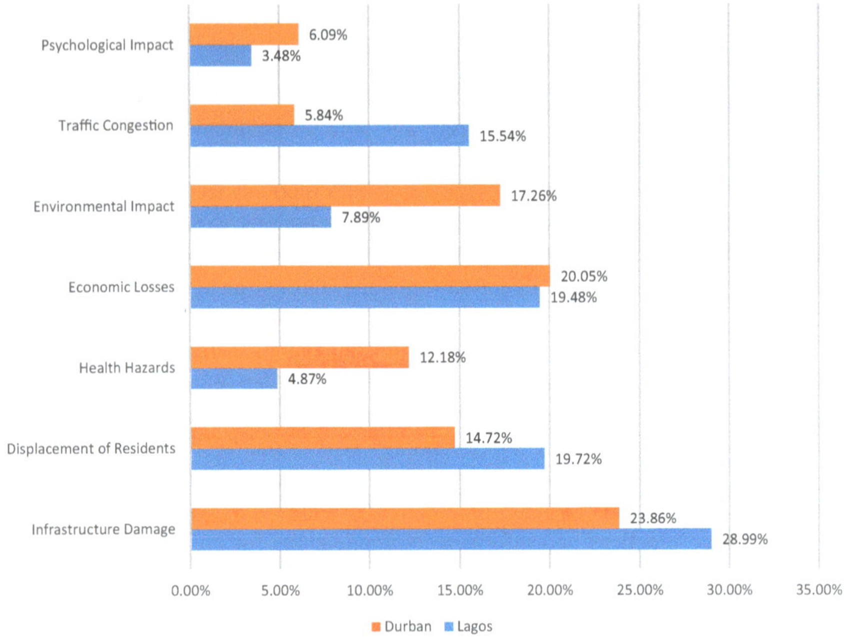

Fig. 2 Cascading impacts of floods in Lagos and Durban

for comprehensive strategies that encompass infrastructure resilience, economic recovery, and environmental preservation. The findings emphasize the urgency of implementing community-level interventions to mitigate the cascading impacts of floods, safeguarding the well-being and sustainability of these African cities.

The investigation into the residents' perceptions of the most affected infrastructure, as detailed in Fig. 3, revealed that the recent floods in the case studies significantly impacted the transportation systems in Lagos (35%) and Durban (36.7%). The residential buildings were also significantly affected, with 32% and 28% of the respondents in Lagos and Durban, respectively, affirming the impacts. These findings highlight the widespread and substantial damage to critical infrastructure, including transportation networks and residential areas, which are integral components of healthy and climate-resilient communities. The disproportionate impact on these vital urban systems underscores the need for targeted resilience measures to protect them, fostering the development of healthier and more climate-adaptive neighborhoods.

Furthermore, the damage to the port of Durban, a key economic asset, demonstrates the far-reaching implications of flood events on regional and national economies, reinforcing the urgency of developing comprehensive, community-level flood mitigation and adaptation strategies to support the overall climate resilience of cities.

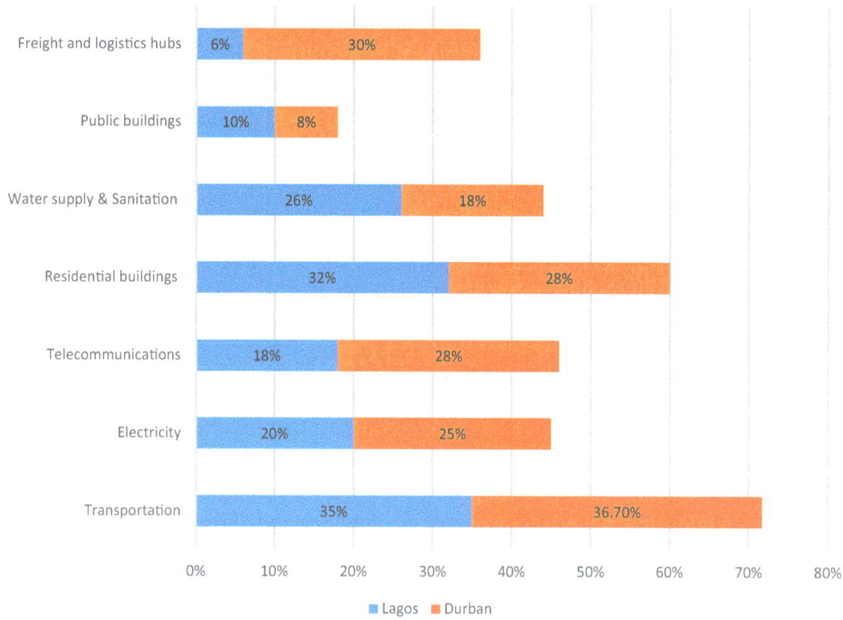

Fig. 3 Residents' perceptions of the most affected infrastructure during recent flood events

3.2 Condition and Capacity of Various Infrastructures to Withstand Frequent Flooding

Qualitative results from stakeholders' perceptions regarding the condition and capacity of infrastructure to withstand frequent flooding events in Lagos and Durban reveal compelling insights. Stakeholders in both cities expressed deep concerns about the vulnerability and limited resilience of the infrastructure in coping with recurring floods.

In Lagos, respondents consistently highlighted the deteriorating state of existing infrastructure, including roads, bridges, and public buildings. They strongly believed that the infrastructure was ill-equipped to handle the increasingly frequent and intense flooding. One local resident shared a personal account, stating:

> Flooding within Lagos communities can be linked to the absence of infrastructure such as drainage, inadequate services such as waste collection, and poor quality housing, most especially in the informal settlements and slums (Interviewee 4: Local resident).

Stakeholders emphasized the urgent need for attention to address these vulnerabilities, as they witnessed firsthand the detrimental impact of floods on infrastructure functionality and longevity. Concerns were raised about the lack of proper drainage systems, inadequate maintenance practices, and outdated design standards that failed to consider the impacts of climate change. These perceptions collectively conveyed a sense of urgency for significant investments and improvements in the infrastructure's resilience to ensure sustainable development in the city.

Similarly, in Durban, stakeholders expressed apprehension about the infrastructure's ability to withstand the impacts of recurring flood events. Specific concerns were raised about the coastal areas, where erosion and inundation posed significant threats to critical infrastructure such as roads, buildings, and utilities. An engineer interviewed highlighted the potential for system failure across the city, stating:

> Durban is now experiencing system failure – where the entire public infrastructure system across the city is likely to collapse over the coming period, and if disasters like floods happen, the impact will be disproportionally damaging in terms of lives lost, damage, and the cost of repairs (Interviewee 15: An engineer).

Respondents stressed the need for innovative engineering solutions, including coastal protection measures and improved drainage systems, to mitigate the risks associated with flooding. Stakeholders also emphasized integrating climate change considerations into infrastructure planning and design processes. They highlighted the necessity of adopting adaptive measures to enhance the infrastructure's capacity to withstand future flood events, considering the projected increase in intensity and frequency.

These findings underscore the urgent need for comprehensive and resilient infrastructure strategies in both Lagos and Durban to effectively address the challenges of frequent flooding. Investments in improved drainage systems, maintenance practices, and climate-resilient design can strengthen the infrastructure's ability to cope with floods and ensure the long-term sustainability of the cities.

3.3 Policy Frameworks and Guidelines Related to Flood-Resilient Infrastructure

The Lagos Resilient Strategy and Lagos Development Plan address the challenges of inadequate infrastructure in Lagos, primarily caused by high population growth and insufficient investment. The strategy recognizes the need for a comprehensive and integrated infrastructure system that can cater to Lagos's current and future residents while also responding to shocks and stresses that affect the city. One key aspect highlighted in the strategy is the development of waterway routes, including canals, through dredging. This measure aims to ensure the safety and unobstructed passage of medium-sized ferries, which can contribute to improved transportation and mobility within the city. By leveraging the extensive water network in Lagos, the strategy seeks to enhance transportation options and alleviate roadway congestion. Another critical component emphasized in the strategy is conducting an energy audit. This audit is essential to assess the energy demand, generation capacity, distribution infrastructure, and metering systems required to provide sustainable and reliable power across the city. The strategy aims to develop and implement measures that improve energy access, efficiency, and resilience by understanding the energy landscape and identifying gaps or inefficiencies.

On the other hand, the Durban climate strategy provides information on several key areas aimed at addressing infrastructure development to enhance climate resilience and promote sustainable development in Durban. These areas include reevaluating and protecting municipal infrastructure in high-risk coastal zones. This approach acknowledges the vulnerability of these areas to climate-related risks such as sea-level rise and coastal storms. Furthermore, the policy recognizes the challenges associated with transportation in a sprawling city like Durban. It highlights the significant greenhouse gas emissions and congestion resulting from inadequate public transport and increased private car usage. To tackle these issues, the strategy mentions developing the Integrated Public Transport Network (IPTN) system called Go! Durban. The aim is to reduce congestion and emissions by promoting densification along transit corridors and encouraging a shift from private to public transport. The strategy also includes the development of guidelines for modifying and retrofitting municipal infrastructure and developments in high-risk coastal zones. This approach recognizes the need to adapt existing infrastructure to the impacts of climate change. Additionally, the strategy acknowledges the importance of ecological infrastructure, which refers to naturally functioning ecosystems that provide valuable services to people. It highlights the role of ecological infrastructure in climate regulation, water provision, flood management, and other services.

Major gaps identified in the policy documents of both cities include the lack of appropriate costing for the proposed infrastructure development and insufficient emphasis on private sector mobilization for funding and development of the critical infrastructure.

3.4 Barriers and Enablers to the Implementation of Flood-Resilient Infrastructure

Findings from the focused group discussions in the two cities regarding the enablers and barriers of implementing resilient infrastructure revealed several key factors. Participants identified common barriers in both cities, including the lack of financial resources and political will and prioritization.

In Lagos, the lack of financial resources was unanimously cited as a significant barrier. The high cost of constructing and maintaining flood-resilient infrastructure is a major challenge, and the city struggles to allocate sufficient funds. Endemic corruption was also highlighted as a major issue, significantly impacting the quality of infrastructure delivery. Poor coordination between government agencies, NGOs, and community groups responsible for implementation has resulted in delays and project completion issues. The survey conducted in the study site indicated that the Lagos state government primarily carried out infrastructure projects, with limited involvement from other stakeholders, hindering effective infrastructure governance.

In Durban, the failure to practice risk-informed urban planning has resulted in poorly designed roads and buildings prone to deterioration, exacerbating the challenges in implementing resilient infrastructure. The lack of data and improved technology was also identified as a major barrier, as access to accurate and up-to-date information is crucial for effective planning and implementation. Furthermore, mistrust and miscommunication between the government and communities, due to the uneven distribution of infrastructure, control, power, and welfare, have contributed to the failure of infrastructure initiatives.

On the positive side, in Lagos, the community's contribution of funds for the construction and maintenance of drainage systems and bridges was highlighted as an enabler. In Durban, the municipality's commitment to addressing flooding challenges, such as conducting topographic mapping and acquiring relevant data, was identified as a facilitating factor. These findings underscore the need for comprehensive strategies that address the financial, governance, and social barriers to implementing flood-resilient infrastructure in Lagos and Durban. Strengthening cross-sectoral collaboration, improving transparency, and fostering community engagement is crucial for enhancing the resilience of urban infrastructure in the face of flood risks.

4 Discussion and Implications

This chapter aimed to identify the cascading impacts of floods on resilient infrastructure in two coastal cities of sub-Saharan Africa, Lagos and Durban. It also explored various policy options and examined the enablers and barriers to implementing flood-resilient infrastructure in these urban centers. The findings shed light on the most prevalent cascading impacts of floods, which include infrastructure damage leading to disruption of critical services, economic losses that strain local budgets, and displacement of residents who lose their homes. Likewise, the World Bank [2] estimates that almost US$200 million per year of infrastructure in coastal cities of sub-Saharan Africa is at risk of flooding, diverting limited resources away from other pressing development priorities.

Furthermore, the study established frequent flooding in these cities has had significant negative impacts. It has disrupted transportation, cutting off access to jobs and essential supplies. It has also rendered residential buildings uninhabitable and disrupted key freight and logistics hubs, hampering the flow of goods and further undermining the local economy. These cascading risks of floods in both Lagos and Durban imply increased loss of jobs and displacement of people in cities already battling severe economic stress, homelessness, and rising poverty levels [18, 22, 23]. This underscores the urgent need to prioritize the development of resilient infrastructure to protect vulnerable populations and maintain economic productivity in the face of worsening flood events.

The chapter also provided insights into the condition and capacity of infrastructure to withstand frequent flooding events in Lagos and Durban. Stakeholders in both cities expressed deep concerns about the vulnerability and limited resilience of existing infrastructure, including roads, bridges, and public buildings. This assertion was supported by previous studies, which found that adaptation to flood in Lagos is relatively low due to a lack of proper drainage systems, inadequate maintenance practices, and outdated design standards that failed to consider the impacts of climate change [24, 25, 26]. Similarly, in Durban, stakeholders expressed apprehension about the infrastructure's ability to withstand the impacts of recurring flood events. These findings underscore the urgent need for comprehensive and resilient infrastructure strategies in Lagos and Durban to address the challenges of frequent flooding effectively.

The identified impacts of flood events on infrastructure, disruption of critical services, economic losses, and displacement of vulnerable populations directly relate to SDG 1 (No Poverty), SDG 3 (Good Health and Well-being), SDG 8 (Decent Work and Economic Growth), and SDG 11 (Sustainable Cities and Communities). The urgent need to prioritize the development of resilient infrastructure highlighted in this study supports SDG 9 (Industry, Innovation, and Infrastructure), which calls for building "resilient infrastructure, promote inclusive and sustainable industrialization, and foster innovation".

The findings from the various policy documents in the case studies revealed that the policy documents recognized the challenges of inadequate infrastructure in Lagos and Durban, as well as the need for a comprehensive and integrated infrastructure system to address both present and future flood occurrences. However, major gaps were identified, including a lack of appropriate costing for the proposed infrastructure development and insufficient emphasis on private sector mobilization for funding and development of the critical infrastructure. This could be attributed to limited technical capacity, a lack of institutional coordination, insufficient data and information, and limited awareness and capacity building [27, 28]. Addressing these gaps and strengthening the policy framework and implementation mechanisms will ensure these cities' successful development and maintenance of flood resilient infrastructure.

The two key barriers common to both Lagos and Durban in the implementation of critical flood-resilient infrastructure are a lack of financial resources and a lack of political will. Insufficient funding not only hinders the construction of necessary flood protection measures, but also jeopardizes the long-term maintenance and upkeep of such infrastructure [29, 30]. This problem is compounded by the diversion of limited public funds towards other priorities, often influenced by corrupt practices and political agendas rather than evidence-based decision-making [31, 32]. The lack of political prioritization of flood resilience projects is a significant obstacle, as it reflects a broader disconnect between government policies and the needs of vulnerable communities [33]. This disconnect can lead to underinvestment in flood mitigation measures, the inadequate enforcement of land use regulations, and the marginalization of the most at-risk populations [34]. The failure to integrate flood risk considerations into urban planning and development processes further compounds these challenges [35].

Poor community engagement and limited public awareness also hinder the effective implementation of flood-resilient infrastructure. When local residents are not adequately informed or involved in the decision-making process, they may be less inclined to support or maintain infrastructure projects, leading to a lack of ownership and sustainability [36]. Additionally, the uneven distribution of infrastructure and the mistrust between the government and affected communities can exacerbate social and spatial inequities, ultimately undermining the overall resilience of these cities [37]. Despite these significant challenges, the examples of community-driven initiatives and data-driven risk assessment in Lagos and Durban demonstrate the potential for local-level solutions to be leveraged and scaled up. However, without addressing the underlying structural and governance issues, implementing flood-resilient infrastructure in these cities will remain a persistent challenge.

Furthermore, the barriers identified, such as the lack of financial resources, political will, community engagement, and governance challenges, mirror the systemic issues that inhibit progress toward many of the SDGs. Addressing these barriers will be crucial for achieving SDG 6 (Clean Water and Sanitation), SDG 13 (Climate Action), and SDG 17 (Partnerships for the Goals), which emphasize the importance of sustainable management of water resources, climate change adaptation, and multi-stakeholder partnerships.

4.1 Towards Participatory Pathways Infrastructure Development

Following the identification of various factors influencing the development of resilient infrastructure for flood risk management in the case studies, this chapter argues that the problems of infrastructure damage, economic losses, displacement, and disruption among others, identified as a result of frequent flooding in African coastal cities, require a pathways approach to build resilient infrastructure that can withstand the cascading impacts of flood events and safeguard the well-being of their communities.

Building upon the concept of climate-resilient pathways discussed in the works of [38, 39, 40, 41], this chapter proposes participatory infrastructure pathways as a means of enhancing the resilience of these cities against future flood incidents. The chapter argues that the most dependable way of proffering sustainable solutions to the various challenges of implementing resilient infrastructure in African coastal cities lies in the adoption of a pathways approach. In this context, pathways refer to the different routes or trajectories that can be taken to promote and enhance the development of flood-resilient infrastructure.

Based on the study findings, the chapter distinguishes four key outcomes for these adaptation pathways (Fig. 4) that promote integrated governance in the provision and management of flood-resilient infrastructure:

- **Promote inclusiveness in the implementation of critical infrastructure:** The pathways should identify and engage actors with different values, goals, and knowledge and proactively ensure access to discussions by different social groups,

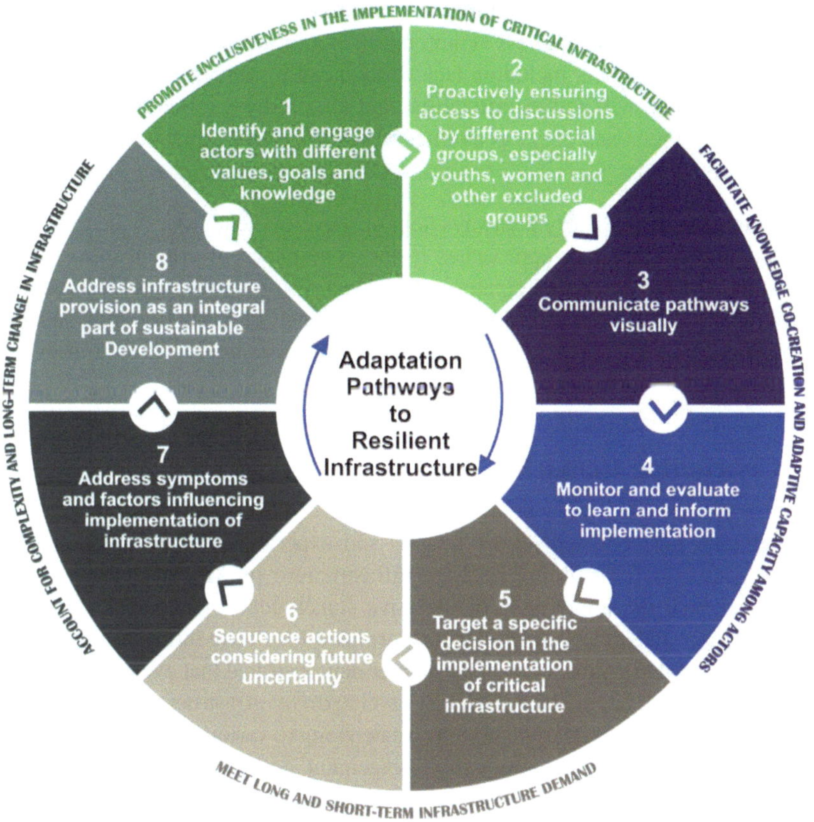

Fig. 4 Participatory adaptation pathways to meet short and long-term infrastructure needs. *Source* Produced by the author based on literature review

especially youths, women, and other marginalized communities, in the planning, design, and implementation of flood resilient infrastructure projects. This helps to align infrastructure solutions with the needs and priorities of all affected groups.

- **Facilitate knowledge co-creation and adaptive capacity among stakeholders:** The pathways should create mechanisms for communicating pathways visually and monitor and evaluate to learn and inform implementation for continuous learning, knowledge exchange, and capacity building among relevant actors, such as policymakers, technical experts, and local community members. This enhances the ability of the system to adapt to changing circumstances and new information.
- **Meet long and short-term infrastructure needs:** The pathways should target specific decisions in the implementation of critical infrastructure and sequencing actions considering future uncertainty. This ensures that communities are protected in the short-term while building resilience for the future.

- **Account for complexity and long-term change in infrastructure development:** The pathways should address symptoms and factors influencing the implementation of infrastructure, including lack of financial resources, community engagement, political will and prioritization, and also address infrastructure provision (including green and grey) as an integral part of the sustainable development of the cities.

Therefore, the participatory pathways help the development process identify, reflect on, and improve existing approaches in the governance of critical infrastructure. The author believes that participatory pathways for infrastructure development will strengthen African coastal cities' resilience against frequent flooding and other extreme events and propel them towards becoming sustainable and safe cities.

5 Concluding Remarks

This chapter has reflected on the lessons and experiences of developing participatory pathways for building resilient infrastructure in African coastal cities. It has highlighted the importance of inclusive stakeholder engagement, co-creation of knowledge, and addressing both short-term and long-term infrastructure needs in a complex and dynamic context. The chapter demonstrated that even though several policies have been put in place at the city level to drive infrastructure development in the cities, their implementation remains weak owing to various challenges, including lack of financial resources, community engagement, political will, and prioritization.

By placing the findings and lessons from this study within the broader framework of the 2030 Sustainable Development Goals, it becomes evident that investment in flood-resilient infrastructure is not only a matter of disaster risk reduction, but also a critical component of sustainable development efforts in coastal cities of sub-Saharan Africa. Addressing the cascading impacts of floods and building the capacity of these urban centers to withstand climate-related shocks will be essential for achieving the Sustainable Development Goals and ensuring the well-being and prosperity of their residents.

Consequently, the chapter emphasized the need for participatory approaches that promote the meaningful participation of diverse actors, visual communication of alternative trajectories, and continuous monitoring and learning. The proposed participatory pathways provide a framework for identifying, evaluating, and improving existing approaches to the governance of critical infrastructure. By promoting inclusive stakeholder engagement, co-creation of knowledge, and addressing both short-term and long-term infrastructure needs, these pathways can enhance the resilience of African coastal cities against frequent flooding and other extreme events, directly supporting the achievement of SDGs 1, 3, 6, 8, 9, 11, 13, and 17.

As cities in Africa continue to grapple with the challenges of rapid urbanization, climate change, and infrastructure deficits, adopting such participatory approaches can help propel them towards becoming smart and safe cities that are better prepared to withstand and adapt to future uncertainties. The insights from this chapter contribute to the growing body of knowledge on collaborative and flexible infrastructure planning in the face of complex socio-environmental challenges, ultimately aligning with the vision of the 2030 Sustainable Development Agenda.

References

1. IPCC (2022) Summary for policymakers [Pörtner H-O, Roberts DC, Poloczanska ES, Mintenbeck K, Tignor M, Alegría A, Craig M, Langsdorf S, Löschke S, Möller V, Okem A (eds)]. In: Climate change 2022: impacts, adaptation, and vulnerability. contribution of working group II to the sixth assessment report of the intergovernmental panel on climate change [Pörtner H-O, Roberts DC, Tignor M, Poloczanska ES, Mintenbeck K, Alegría A, Craig M, Langsdorf S, Löschke S, Möller V, Okem A, Rama B (eds)]. Cambridge University Press, Cambridge, UK and New York, NY, USA, pp 3–33. https://doi.org/10.1017/9781009325844.001
2. The World Bank (2022) Living on the water's edge. Flood risk and resilience of coastal cities in Sub-Saharan Africa. https://documents1.worldbank.org/curated/en/099042623122580007/pdf/P17703604875c205e0a95b0d3f8997b32c8.pdf
3. Botzen WW, Deschenes O, Sanders M (2019) The economic impacts of natural disasters: a review of models and empirical studies. Rev Environ Econ Policy 2019:1
4. Carrera L, Standardi G, Bosello F, Mysiak J (2015) Assessing direct and indirect economic impacts of a flood event through the integration of spatial and computable general equilibrium modelling. Environ Model Softw 63:109–122
5. Mguni P, Herslund L, Jensen MB (2016) Sustainable urban drainage systems: examining the potential for green infrastructure-based stormwater management for Sub-Saharan cities. Nat Hazards 82(2):241–257
6. World Economic Forum Boston Consulting Group (BCG) (2018) Towards a reskilling revolution: a future of jobs for all. In: World economic forum
7. Magidi M (2018) Sustaining livelihoods in Norton town after deindustrialisation in Zimbabwe. Doctoral dissertation, University of KwaZulu-Natal, Durban
8. Okunola OH, Olawuni PO (2022) Determinants of household flood resilience strategies in Kaduna Metropolis, Nigeria. Urban Clim 41:101216
9. Mensah H, Amponsah O, Opoku P, Ahadzie DK, Takyi SA (2021) Resilience to climate change in Ghanaian cities and its implications for urban policy and planning. SN Soc Sci 1(5):118
10. Cradock-Henry NA, Blackett P, Hall M, Johnstone P, Teixeira E, Wreford A (2020) Climate adaptation pathways for agriculture: insights from a participatory process. Environ Sci Policy 107:66–79
11. Maru YT, Smith MS, Sparrow A, Pinho PF, Dube OP (2014) A linked vulnerability and resilience framework for adaptation pathways in remote disadvantaged communities. Glob Environ Chang 28:337–350
12. Trejo D, Gardoni P (2023) Special issue on adaptive pathways for resilient infrastructure: an introduction. Sustain Resilient Infrastruct 8(sup1):1–2. https://doi.org/10.1080/23789689.2022.2139564
13. Okunola OH (2024) Stakeholder engagement in disaster recovery: insights into roles and power dynamics from the Ahr Valley, Germany. Int J Disaster Risk Reduction 114:104960. https://doi.org/10.1016/j.ijdrr.2024.104960
14. Mulligan J, Bukachi V, Clause JC, Jewell R, Kirimi F, Odbert C (2020) Hybrid infrastructures, hybrid governance: new evidence from Nairobi (Kenya) on green–blue–grey infrastructure in informal settlements. Anthropocene 29:100227
15. Hayes S, Desha C, Burke M, Gibbs M, Chester M (2019) Leveraging socio-ecological resilience theory to build climate resilience in transport infrastructure. Transp Rev 39(5):677–699

16. Roy R, Gain AK, Hurlbert MA, Samat N, Tan ML, Chan NW (2021) Designing adaptation pathways for flood-affected households in Bangladesh. Environ Dev Sustain 23(4):5386–5410
17. Zuniga-Teran AA, Gerlak AK, Mayer B, Evans TP, Lansey KE (2020) Urban resilience and green infrastructure systems: towards a multidimensional evaluation. Curr Opin Environ Sustain 44:42–47
18. Okunola OH, Simatele MD, Daramola O (2023) To live or to die: cultural and social factors influencing flood preparedness in Nigerian cities. Afr Geograph Rev 2023:1–15
19. Okunola OH (2022) How policy can address frequent flooding in African coastal cities. Africa at LSE. https://blogs.lse.ac.uk/africaatlse/2022/05/09/how-policy-can-address-frequent-flooding-in-african-coastal-cities-climate-change/
20. Global Facility for Disaster Reduction and Recovery (2024). https://thinkhazard.org/en/report/77311-south-africa-kwazulu-natal
21. Durban Climate Change Strategy (2022). https://www.durban.gov.za/storage/Documents/Climate/DCCS_Strategy.pdf
22. United Nations, Department of Economic and Social Affairs, Population Division (2019) World population prospects 2019: highlights. New York (US): United Nations Department for Economic and Social Affairs, 11(1):125
23. Okunola OH, Balogun K, Ojo DB (2023) Indigenous knowledge and flood resilience strategies in African coastal cities: from practice to policy. Water Environ Sustain: Case Stud Developing Countries 161–172.
24. Douglas I, Alam K, Maghenda M, Mcdonnell Y, McLean L, Campbell J (2012) Unjust waters: climate change, flooding and the urban poor in Africa. In: Adapting cities to climate change. Routledge, London, pp 201–223
25. Lawanson OI, Proverbs D, Ibrahim RL (2023) The impact of flooding on poor communities in Lagos State, Nigeria: the case of the Makoko urban settlement. J Flood Risk Manag 16(1):e12838
26. Okunola OH (2020) Residents vulnerability analysis and explanation of development-induced disasters in a Sub-Saharan African city. Int J Dis Resilience Built Environ 11(1):71–84
27. Adom RK, Simatele MD, Reid M (2022) Addressing the challenges of water-energy-food nexus programme in the context of sustainable development and climate change in South Africa. J Water Clim Change 13(7):2761–2779
28. Shackleton S, Ziervogel G, Sallu S, Gill T, Tschakert P (2015) Why is socially-just climate change adaptation in sub-Saharan Africa so challenging? A review of barriers identified from empirical cases. Wiley Interdiscip Rev Clim Change 6(3):321–344
29. Amoako C, Frimpong Boamah E (2015) The three-dimensional causes of flooding in Accra, Ghana. Int J Urban Sustain Develop 7(1):109–129
30. Ziervogel G, Cowen A, Ziniades J (2016) Moving from adaptive to transformative capacity: building foundations for inclusive, thriving, and regenerative urban settlements. Sustainability 8(9):955
31. Transparency International (2019) Corruption perceptions index 2019. https://www.transparency.org/en/cpi/2019
32. Okunola OH (2025) Exploring multi-level governance arrangements in disaster recovery: a study of lagos, Nigeria. Int J Disaster Risk Reduction 105254.https://doi.org/10.1016/j.ijdrr.2025.105254
33. Adelekan IO (2010) Vulnerability of poor urban coastal communities to flooding in Lagos, Nigeria. Environ Urban 22(2):433–450
34. Ajibade I, McBean G (2014) Climate extremes and housing rights: a political ecology of impacts, early warning and adaptation constraints in Lagos slum communities. Geoforum 55:76–86
35. Anguelovski I, Shi L, Chu E, Gallagher D, Goh K, Lamb Z, Teicher H (2016) Equity impacts of urban land use planning for climate adaptation: critical perspectives from the global north and south. J Plan Educ Res 36(3):333–348
36. Frick-Trzebitzky F, Baghel R, Bruns A (2017) Institutional bricolage and the production of vulnerability to floods in an urban informal settlement, Kolkata. Environ Plan E Nat Space 1(4):612–633

37. Ziervogel G, Frayne B (2011) Climate change and food security in Southern African cities. Urban Food Secur Ser 8:1
38. Brown V, Lambert J (2015) Transformational learning: Are we all playing the same 'game'? J Transform Learn 3(1):35–41
39. Butler JRA, Wise RM, Meharg S, Peterson N, Bohensky EL, Lipsett-Moore G, Dunstan P (2022) 'Walking along with development': climate resilient pathways for political resource curses. Environ Sci Policy 128:228–241
40. Werners SE, Wise RM, Butler JR, Totin E, Vincent K (2021) Adaptation pathways: a review of approaches and a learning framework. Environ Sci Policy 116:266–275
41. Okunola OH, Werners SE (2024) A multi-dimensional framework for assessing disaster recovery pathways: lessons and experiences from Germany and Nigeria. Int J Disaster Risk Reduction 112:104777. https://doi.org/10.1016/j.ijdrr.2024.104777

Healthy Building and Resilient City in Dezful: A Study of Passive Climatic Solutions

Amir Tayyebi and Kourosh Momeni

Abstract The ancient city of Dezful, with a history of nearly 5000 years, is located in Khuzestan province in the Middle East. This region has a hot and semi-humid climate, and its summers are hot and humid and its winters are cool and rainy. So much so that today, people in this area cannot spend the summer without the help of air conditioning devices. The air temperature, with an average of 49 degrees Celsius in summer, turns all the green fields of this area into yellow fields. But in the past, people in this area were able to create suitable conditions for living in the hot summer months by using natural and creative solutions. This city is located on the banks of the Dez River. Creating corridors and underground alleys called Qomesh from the old houses to the banks of the Dez River, which brought the cool air of the river to the residential houses through underground channels, and also a way to reach the river from the depths of the earth without being exposed to direct sunlight. Also, the use of hand-made caves called Mahali Kat in the coastal walls of the Dez River was a place for families to take refuge by the river and stay long in the hot summer. Shavadun was also one of the hand-made underground spaces that were ventilated by a small fan. All the above shows that the people of Dezful City in the past, by using passive climatic solutions, provided the flexibility of Dezful city for living in the exhausting summer conditions. Solutions that continue their lives as a recreational tourist attraction and still the people of Dezful use them.

Keywords Passive climatic solutions · Dezful · Urban resilience · Healthy building · Qomesh

1 Introduction

Dezful is a city of almost 5 millennia old history, located in the southwest of Iran Khuzestan province. Dezful is a significant settlement for its prime location on the high left bank of Dez River, near the Ahwaz region road descending from Karun

A. Tayyebi (✉) · K. Momeni
Architecture and Planning Department, Jundi-Shapur University of Technology, Dezful, Iran
e-mail: a_tayyebi@jsu.ac.ir

A. Cheshmehzangi et al. (eds.), *Designing Healthy Cities*, Urban Sustainability,
https://doi.org/10.1007/978-981-96-5538-0_8

and uppermost navigable stretches in general part—below ancient outfalls to pursuit across river marshes also extensive guide northwestward districts notable irrigation irrigated by any earlier than politically quite suggestive gateways upriver lowering chance grievously [1]. The name Dez-fool has come from the formances of two words in Farsi/Persian "diz" meaning fortress and 'pul' for bridge, thus highlighting its historical link to Bridge as a fortified—limited access bridge. Here is the mighty bridge over the Dez River, constructed in Sassanian times by Shapur I from Roman prisoners of war—a reminder that this was once an advanced city [1]. This legacy bridge continues to be a significant landmark in history.

Dezful is located 650 km northwest of Tehran and 155 km south of Ahvaz, the provincial capital. This city has summer which is scorching hot and winter warm with most of its rain from November to April. However, these difficult climatic conditions did not prevent Dezful from being the center of civilization as it utilized several passive method to bring comfort in their habitable environment. The study of passive climactic solutions has no choice but to play a major part in current urban planning, if we intend on developing things sustainably. At a time when cities around the world are grappling with both rapid urbanization and climate change, these lessons of ancient practice can shed light on contemporary challenges. This chapter examines how Dezful, best known for its history and sustainable yet traditional passive climatic solutions can be seen as a step towards healthy buildings and thus resilient cities today.

Benefits of Passive Climatic Solutions Use natural processes and materials for temperature control helps to reduce the need in energy. Thay are a better investment. This solutions improve not only architectural appearance but also health. People can physically interact with passive systems. As passive solutions reduce the need for mechanical cooling systems, they save a lot of energy. Strategies such as natural ventilation, shading and thermal mass allows a maximum comfort with the minimum energy input.

This would also have an environmental impact by generating a decrease in the greenhouse gases caused by reduced energy demand and industry resulting in helping mitigate or stop the urban heat island effect, which is very important for global warming. These solutions can lead to an even more sustainable urban environment. Constructing a building with a passive design can see an overall reduction in the cost required to build and maintain the structure over its lifespan. These solutions can offer a great return on investment for energy bills and maintenance cost savings for new constructions and retrofits. Passive solutions provide more even temperature bands and better air quality, so everyone from family to friends will love spending time inside. Healthy environments for living and working are provided through natural ventilation and daylight and it will mean that people feel good in a building.

When passive design elements are incorporated into contemporary architecture, it helps the cultural heritage to survive plus an added aesthetic value [2]. It helps give a sense of place and connection to the past which enhances our urban experience. The buildings are designed to be more resilient, and they take the help of the passive solutions for climate. This represents a sustainable replacement for ubiquitous and noisy air-cooling apparatus on roadways challenged by global warming of uncertain destiny.

These passive climatic solutions implemented in Dezful shall be discussed and their potential importance for contemporary urban planning examined. These strategies help us not only tackle short-term issues around energy consumption and climate change, but also build a more resilient cities that are healthier to live in while fostering innovation. Following this thread, we can extract many lessons that could be applied to contemporary urban centers in order to secure a sustainable city and region for generations.

2 Historical Background

Dezful, a city of Dez River was the capital of Bakhtiaris in the 16th to early eighteenth centuries this long history shows that it has been one of the most ancient human dwelling places [3]. The rest of the earlier city has been placed at a number of points across an archaeological landscape with remains from ancient Sumer and Elam (specifically, Tappeh Anbar). Dezful is located near the ancient Lullubi (Lulubian) city of Sadak [4]. According to archaeological findings, the history of Dezful can be traced back for thousands of years, making it one of the ancient cities in Iran. Its name is thought to originate from the Pahlavi language, giving it the meaning "two bridges," and represents its strategic value as well as how well connected it was.

The most important period of Dezful is related to its history during the Sassanian Empire [1]. The bridge over the Dez River was constructed in this period, and historians now consider it an important expression of military engineering. It is also called the Dezful Bridge, proof of how much an advanced level of engineering Iran possessed in the ancient era and one more historical site that makes it worth visiting. After the Muslim conquest of Persia, Dezful prospered as an important regional administrative center. Dezhpul was one of the founders of what is now called Dezful, however with linguistic differences in Arabic, Dehz-well turned to Dez. During the Islamic era, Dezful continued its significance as a base for trade and culture in the area.

Dezful came to architectural and cultural milestones during the Qajar era. Its classical Perso-Islamic architecture dates back to the Qajar era, featuring traditional houses with windcatchers and alcoves adorned by elaborate tile work [1]. Historically significant, these buildings are part of Dezful's deep-rooted culture and showcase the past through time. Dezful is a city with the influence of ethnics like Arabs, Bakhtiaris, and Persians. This cultural diversity also shows up in the local accent, customs, and arts. Festivals and celebrations like Nowruz (Persian New Year) further illustrate the city's colorful cultural life. Apart from the Dezful Bridge, several historical sites are located in this city. The rich cultural and historical backdrop along with the strategic location, made Dezful into a unique gem in Iran. An important knowledge to comprehend in order to experience the city's forward-thinking passive climatic solution and its relevance up until today's urban planning.

3 Methodology

In this study, efforts are going to be made through the following main approaches in order to such solutions in Dezful. First is historical analysis (Analysis of historical texts and landmarks within architecture and archaeological findings). According to geophysical excursions that have happened so far. Second is field Surveys, including site visits and documentation of existing passive climatic structures like Qomesh Corridors, Mahali Kat Caves, and Shavaduns. This includes photo evidence and architectural plans.

4 Climate of Dezful

4.1 *Description of Dezful's Hot and Semi-Humid Climate*

Dezful is set in the extreme heat zone of semi-arid climate (Köppen classification) just like most of southwest Iran, with scorching and long summer heat peaks over 50 °C [5]. Dezful summers are exceptionally hot, routinely exceeding 46 °C (114.8°F) in July and August. In these months, rain is near zero, and humidity levels decrease significantly to average around 22–32%2. High Heat + Low Humidity = HARSH summers [5]. Conversely, the winters in Dezful are mild; with temperatures hovering around 5.321 °C (41.3798 °F) for its coldest months, to around highs of about 20°3 cacius (68°F) [5]. One of the rainy seasons, especially from January through March with a peak in January which is also one of the wettest months receiving 101 mm (3.98 inches) on average [5].

In general, the city of Dezful has a very hot summer and mild winters with freezing nights (in winter) causing frost. This causes vulnerabilities for crops when subjected to intense drought in December and severe heat waves in July/August, even though they can rarely be seen two months apart.

4.2 *Seasonal Variations and Their Impact on Living Conditions*

Dezful has a dramatic contrast between its most intensely hot summers and woefully frigid winters, resulting in very difficult living conditions for those residing there. The city averages 114.8F during the summer, with frequent temperatures higher than from May to September (> 46 °C) [5]. The temperatures can easily exceed levels that are tolerable, and between this reached near balmy 20 C degrees the day before yesterday. Low humidity also does not make it any easier. People resort to using air conditioners and other cooling methods for long durations due to the unbearable heat. A greater incidence of sunstroke and dehydration but not least a serious hindrance on any outdoor activity or exercise during the hottest months. The other issue is that during this time there will not be any rainfall, so water availability becomes a real problem, and it requires good management of the same.

Winter is much different with semi-warm temperatures of about 5.3 °C (41.5°F) and 20 °C [5]. It is the season that gives more comfort to residents offering a greater variety of outdoor activities. Higher winter precipitation, specifically in January to refill water sources and benefit farming operations. But the cold will still be an issue, particularly for those without heat. The seasons of Spring and Autumn mark the transition periods during which temperatures are moderate, with more equitable levels of humidity. This allows residents to enjoy a greater quality of life due to being able to take part in activities and farming practices which would be unfeasible if it were too cold.

On the whole, seasonal variations in Dezful have profound impacts on daily life ranging from cooling and heating strategies to natural resources management practices; such as agriculture activities or water distribution. Likewise, the divergence has important implications for how cooling demands in these regions are addressed to ensure that residents can adapt and live comfortably.

5 Traditional Passive Climatic Solutions

5.1 *Qomesh: An Ingenious Water Management System*

This is a system of one-way qanats built about four or five centuries ago that used the Dez River as its origin to store and transport water into the urban area located in Dezful [6]. Discovered in Shushtar, it is called "Softeh" and its antiquity dates back to events such as the Sassanid kings.

The 13–16 m elevation drop from the city down to the river is what led them to mastermind a very accurate and ingenious calculation known as Qomesh. The system was devised to deliver the greatest possible pressure required for a 30–40 km-long underground route while honouring seasonal discharge in an environment where river water levels are at their lowest during the winter and highest in summer [3]. Some of the important Qomesh systems in this city: Hajiyoun, Sarkandeh, Aghamir, and Choqabafon Sarbatagh. The port of Qomesh is called "Sarbatagh" (Figs. 1, 2, 3).

Although these Qomesh systems work just like Qanats as well, they are intended to transport water from rivers and streams in contrast to collecting ground water. Dezful is also home to about 15 Qomesh systems that were active decades after the first widespread use of kahrizes in its industrial age between the early discreet stages until well into the mid-late oil engine times from around the late-1950s [3]. However, due to urban development and the spread of new water supply networks, irrigation wells have gone out of use so they are abandoned. The Qomesh system of Choqabafon is open to the public and awaiting people's visit.

Fig. 1 The entrance of a Qomesh (Sarbatagh) (photo taken by the authors)

Fig. 2 Access stairs to Qomesh (photo taken by the authors)

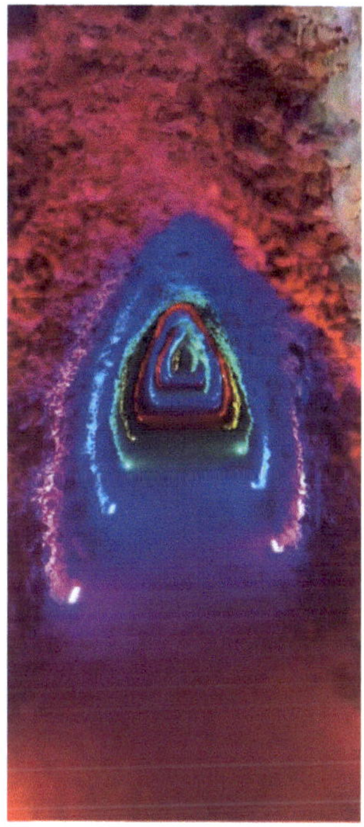

Fig. 3 Qomsh water transfer route (photo taken by the authors)

5.2 *Mahali Kat Caves*

Iran is a brilliant ancient illustration of passive climatic solutions that were implemented [7]. But these really like caves, or better say the man-made holes have been etched into banks alongside Dez Road. These caves were built hundreds of years ago which suggests that the locals might know a thing or two about keeping things cool with minimal resources. The making of these caves involved intricate rock excavation, ensuring the inside was cool and could be inhabited. The caves were placed for maximum advantage of the nature around, and used the closeness to cool through proper positioning near rivers. The thick stone walls did a good enough job to keep the insides of those buildings well conditioned regardless of external weather conditions (Figs. 4, 5).

The main benefit of the Mahali Kat Caves is that they provide a cool highlight in the heat-stressed summer. The rock insulates the caves and keeps them naturally cool, plus their location adjacent to a river makes for an even cooler microclimate inside that is much colder than outside. The caves are an effective way of providing passive cooling capacity to buildings and during the hot summers can help save on powered air conditioning costs. This saves energy, as well as prevents the depletion of valuable resources from recovering. The caves draw in the cold air from the river whereas thick rock walls help prevent heat enter inside. It is an eco-system that keeps the caves full of pure, fresh air and cold like heaven in scorching summer.

Fig. 4 A view of the caves in the heart of the wall of Dez river (Kat) (photo taken by the authors)

Fig. 5 Schematic section of Dezful Kat cave (photo made by the authors)

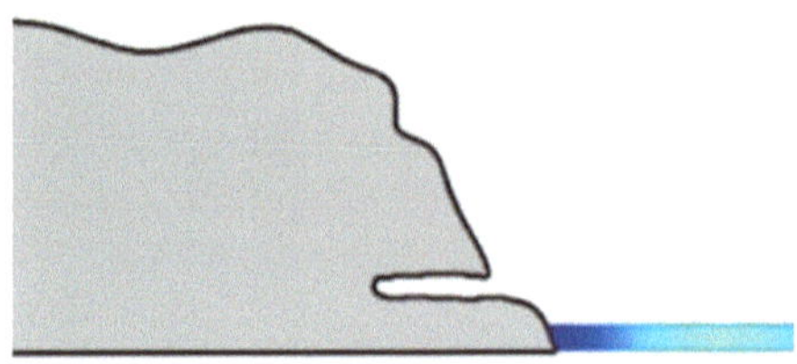

5.3 *Shavadun Spaces*

These are subterranean chambers that create Shavadun spaces, a specific feature of the traditional construction in Dezful houses that their wind-catchers allow for natural cooling throughout summer months at the shallow depth they have been there. Usually, they are 5–12 m underground; in this way, the temperature of several hundred million years ago is taken advantage of offering a pleasant living space. Shavaduns rely on a unique ventilation technique that combines natural airflow with thermal mass cooling [8]. A highlight of the Shavadun ventilation was Darizeh, vertical shafts that were connected to underground spaces through them and brought air from the surface. The shafts pull cooler air in at the base and exhaust excess heat out while allowing it to passively move air due to convection. This movement also called the stack effect, ensures a steady turnover of fresh air to keep occupiers comfortable within (Fig. 6).

In addition, the massive earthen walls of Shavadun act as thermal mass with a high heat capacity that absorb solar radiation during day time and release it at night. This natural insulation helps to trap in temps, keeping the interior cooler during summer and warmer throughout winter. Especially in the hot subtropic and semihumid climate of Dezful, the design has been contributed largely to natural thermal comfort inside houses. Utilizing the stable heat of underground, they can keep inside temperature steady at 21–24 degrees Celsius all year long [9]. A much cooler option than summer's

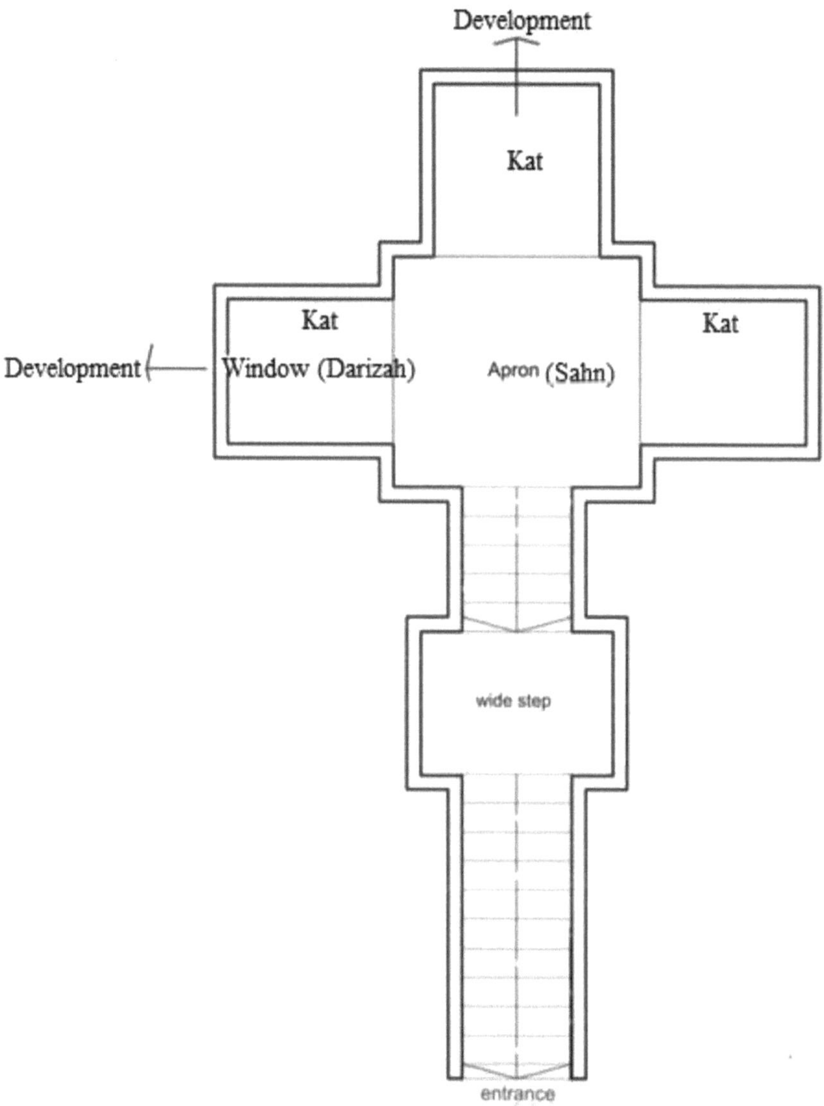

Fig. 6 General plan of Shavadun (adapted by the authors from the original *Source* [9])

outside temperatures, offering a natural cooling solution to save on energy. Natural ventilation, combined with cooling from thermal mass as well occupants inside of the house and other passive systems completely eliminate artificial air conditioning allowing for low energy use living environment [9]. The sensor-equipped Darizeh shafts of the Qanat system passively draw cool air inside the Shavadun, which in addition to being cooled gets purified and its moisture reduced (Fig. 7).

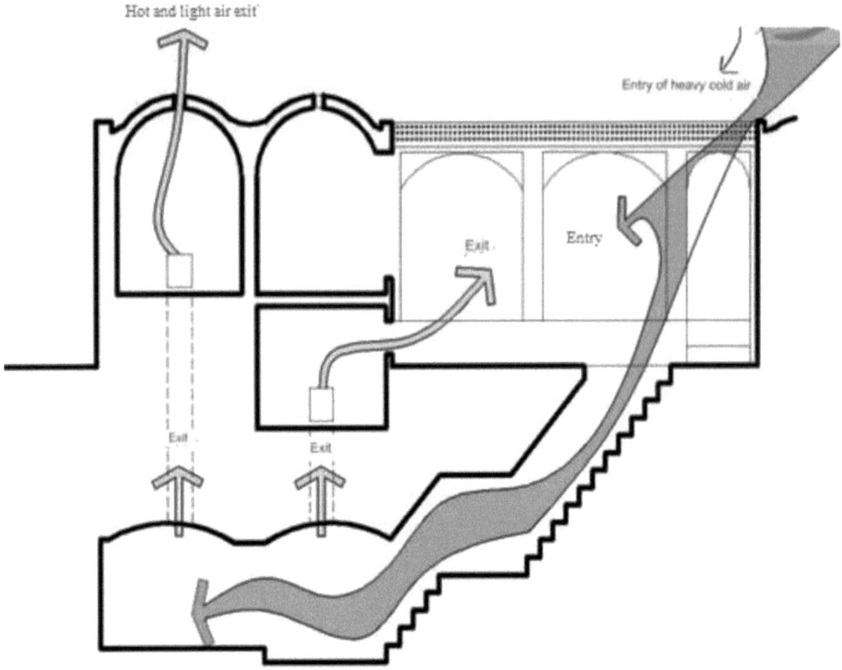

Fig. 7 A general section of the Shavadun, how the cold night air enters and exits heated air (Adapted by the authors from the original source [9])

To sum up, Shavadun spaces are a good example of the utilization of passive climatic solutions in traditional architecture. Their inventive use of ventilation and ability to regulate interior climates along with their novel perspective on the combination of traditional knowledge and sustainable technologies for this purpose can be used in hot regions around the globe.

6 Tourism and Cultural Heritage

These facilities (Qomesh corridors, Mahali Kat caves, and the Shavadun subsurface spaces) are tended to by a small crew who maintain them in working order while still allowing physical access. Their likeness in style and function is what allows preservation to be achieved with consistent upkeep, maintained restoration efforts, and the support of traditional materials. These activities have doubled not just to ensure these structures but also the continuance of knowledge and skills they had in its construction, manufacturing, or setting up.

The passive climatic strategies used in the vernacular architecture of Dezful have made these houses an interesting challenge for tourists [10]. People come here from all over to see what the locals did in order to battle with living up. Many guided tours emphasise where air runs from Qomesh corridors along the Dez River into living spaces, and hidden spots like caves Mahali Kat or rooms Shavadu help people to escape the summer heat [11]. More than educational though, the tours also celebrate both past and contemporary innovation in utilizing our finite resources to meet needs while preserving them for future generations. Being able to see these passive cooling methods is a rare treat and makes Dezful a must-visit destination for those who enjoy historical architecture and sustainable living techniques [11].

Today tourists in Dezful find a harmony between these historical trends and modernity. This will help to immerse tourists in the use of passive climatic solutions by turning traditional structures into widespread materials used for culture, museums, and guesthouses. Sleeping in a working Qomesh hall or exploring a Mahali Kat cave can create rare and vivid impressions. Moreover, the city's heritage will be promoted through its hosting of cultural festivals and events serving to expose traditional modes of construction and passive cooling strategies as an additional tourist draw. These events allow tourists to more deeply engage with traditional culture and history, increasing their admiration for the generation of entrepreneurs who created innovative solutions in the city.

7 Sustainability and Resilience

Dezful, a small ancient town, exemplifies how the use of traditional solutions can used to live because we need it sustainable. The application of passive climatic technologies like Qanat, Mahali Kat and Shavadun depicts the intelligence of our ancestors to construct sustainable living environments without or with minimal support from technology. These practices used natural resources and these were interwoven with climatic conditions to be above the comfortable living conditions of humans within by manipulating what was occurring outside based on understanding factors in local climate [11]. Qomesh benefited from its underground alleys and corridors, which made it possible to access the cold air that descended off the Dez River. Residents could then benefit from a natural cooling effect in their homes with the passage of this air that had been channeled underground. In addition to that, this method made air conditioning almost a non-requirement and hence reduced overall energy use, which led humans towards sustainable living.

Hand-hewn caves along the riverbank of Mahali Kat offered a cool respite from summer. Carved into the coastal walls, these caves offered the family a cool and shaded escape from sizzling temperatures. Building a structure of the earth using natural materials and situating these caves where they belong celebrates job creation that is ecologically, and economically feasible for the long run. Shavadun, the underground spaces with little ceiling fans positioned in some areas, evince even more clearly how Dezful's inhabitants have used every source possible. This helped facilitate fresh air circulation and to help maintain a comfortable temperature, reducing the need for a mechanical cooling system [10]. Shavadun shows the possibility of achieving sustainable thermal comfort using passive solutions, by integrating natural ventilation techniques.

What Dezful and its predecessors offer are lessons for contemporary urban planning in a world where carbon reduction will reshape the human environment. As cities the world over wrestle with how to beat back climate change and meet increasingly voracious energy demands, this review suggests that integrating long-practiced ways of thinking about links between matter and form can help ensure environmentally sustainable city-shaping practices capable of providing resilient solutions. Qomesh can serve as a model for urban planners, showcasing that modern buildings should be designed to include an underground cooling system. Natural air flows and underground heat exchangers can cost-effectively replace much of the air-conditioning by providing plenty of cool, fresh-conditioned space for a fraction of today's high energy consumption and gas emissions. This might be especially useful for hot and humid climates with high cooling demands.

This model can also be replicated while designing buildings to foster sustainable living. This can enhance thermal comfort and aim to reduce the urban heat island effect, through creating more shaded cooler spaces in an otherwise dense city environment. These principles can be applied to all public spaces, housing, and commercial development individually so we create a more resilient urban environment. Shavadun-inspired ventilation strategies may inform contemporary architecture. Natural ventilation and airflow have the potential to keep indoor spaces cool without requiring a significant amount of energy. This is achieved by the careful placement of windows, vents, and other significant architectural elements that create ideal cross-ventilation across the property.

8 Conclusion

The city Dezful, which comes from one of the ancient civilizations in Iran has a difficult climate to live with; here people created passive solutions for climatic problems. Historical contexts featuring underground habitable spaces such as the Qomesh corridors, Mahali Kat caves, and Shavadun had historically allowed residents to beat the unprecedented summer heat. Utilising natural ventilation, shade as well as subterranean coolness the inhabitants of Dezful succeeded in dramatically reducing several climatic factors without resorting to contemporary air-conditioning technology. These age-old practices offered solace and evidenced mature responsible living in harmony with nature.

As the world urbanized at an unprecedented rate and climate change began to reveal its impacts in earnest, Dezful history has a great deal left to teach us. Its strategies may offer important lessons that urban planners and architects working to build sustainable, resilient cities can gain inspiration from. The implementation of passive climatic solutions into modern building trays leads to a reduction in energy consumption as well as greenhouse gas emissions while also increasing indoor comfort. Since most such methods are processes being forgotten, wouldn't it be possible to improve and tweak these traditional approaches in service of contemporary needs? New establishments and retrofits to existing buildings will be greener, and more energy efficient following the age-old instructions and fusing them with modern technological developments. Even more importantly, these solutions can act as a blueprint for other areas with commensurate climatic challenges to pivot the wider world towards sustainable urban development.

While the passive climatic solutions of Dezful do not quite mitigate this idea, they nevertheless illuminate a historical level of resilience in city planning which can inspire cities to find resonance with their past as well as unlock potential avenues for future development. The adoption of these strategies for healthier buildings and more resilient cities provides an ecological balance that accommodates human activity as one with nature while fostering the growth of generations to come in a sustainable, livable environment.

References

1. Abdollahi M (2021) Dezful over time. Yahdgare Andishe, Tehran (**in Persain**)
2. Dumitraşcu AI, Corduban CG, Hapurne TM, Bliuc I (2022) Passive house design in temperate continental climate for increasing indoor comfort conditions. In: IOP conference series: earth and environmental science. IOP Publishing, New York
3. Zalaghi Y, Esmaili-Jelodar ME, Khosrozadeh A (2024) Supplying water to the Sasanian-Islamic city of gundishapur from the River Qanat of Dezful City: design a hypothesis. Parseh J Archaeol Stud 27:171–195 (**in Persain**)
4. Frye RN (1984) The history of ancient Iran. C.H. Beck, New York
5. Imam Ahvazi MA (2003) Geographical history of Dezful. Darolmomenin, Dezful (**in Persain**)
6. Alemohammad SH, Gharari S (2017) Qanat: an ancient invention for water management in Iran. In: Found in proceedings of water history conference, Delft, The Netherlands
7. Izadpanahi P, Farahani LM, Nikpey R (2021) Lessons from sustainable and vernacular passive cooling strategies used in traditional Iranian houses. J Sustain Res 3(3):1
8. Mohammadshahi S, Nili-Ahmadabadi M, Nematollahi O (2016) Improvement of ventilation and heat transfer in Shavadoon via numerical simulation: a traditional HVAC system. Renew Energy 96:295–304
9. Bina M (2008) Climatic analysis of shavadons in Dezful houses. Honarhayeh Ziba 33:37–46 (**in Persain**)
10. Mohammad Alinezhad F (2019) Passive cooling in Shavadoon of traditional buildings of Dezful city: cooling through renewable energy sources. Iran J Energy Environ 10:115–120
11. Sadoughi A, Kibert C, Sadeghi FM, Jafari S (2019) Thermal performance analysis of a traditional passive cooling system in Dezful, Iran. Tunnel Undergr Space Technol 83:291–302

Leveraging Metro Data to Analyze Population Mobility and Reduce Epidemic Risk in Megacities: A Novel Approach for Healthier City Development in Shenzhen, China

Rongpeng Zhang, Sheng Jiao, Ziyi Lin, Yue Hu, and Wenke Zong

Abstract The interplay between population mobility and epidemic risk is critical for fostering resilient urban environments, especially in rapidly growing megacities. This chapter introduces a novel approach for healthier city development by utilizing metro data to analyze how population mobility influences the spread of infectious diseases. By leveraging the extensive information provided by metro systems, cities can gain valuable insights into epidemic dynamics, leading to more effective strategies for risk mitigation and public health promotion. While population mobility drives economic growth and urban development, it also presents significant public health challenges. The rapid movement within densely populated areas can create conditions conducive to disease transmission. Modern metro systems, which enhance connectivity, may inadvertently facilitate disease spread due to high passenger volumes and frequent interactions. The COVID-19 pandemic highlighted the unique challenges faced by megacities with extensive metro networks in epidemic prevention. Traditional approaches often do not account for the complex dynamics of population mobility within these systems. Focusing on high-risk areas in Shenzhen, known for its extensive metro network, this chapter employs a case study approach, using statistical and spatial analysis methods to identify patterns and develop targeted strategies. By integrating transportation data with public health measures, this chapter presents a framework that megacities can adopt to enhance epidemic preparedness and resilience. Combining these insights with innovative urban planning, the chapter provides a pathway for building healthier, more resilient cities that prioritize public health. It offers guidance for future city planning and public health initiatives, aligning with the broader theme of sustainable and healthy urban environments.

Keywords Megacity · Healthy city · Epidemic risk · Urban resilience · Population mobility · Metro data

R. Zhang (✉) · S. Jiao · Z. Lin · Y. Hu · W. Zong
School of Architecture and Planning, Hunan University, Changsha, People's Republic of China
e-mail: zhangrongpeng@hnu.edu.cn

A. Cheshmehzangi et al. (eds.), *Designing Healthy Cities*, Urban Sustainability,
https://doi.org/10.1007/978-981-96-5538-0_9

1 Introduction: Leveraging Metro Data for Healthier City Development

1.1 Urbanization and Public Health Challenges

Rapid urbanization has been a defining trend of the twenty-first century, with more than half of the world's population now living in urban areas. This shift is particularly pronounced in developing regions where cities like Lagos, Mumbai, and Shenzhen are expanding at unprecedented rates. Urbanization brings significant benefits, such as economic growth, increased access to services, and improved infrastructure. However, it also poses considerable challenges. As cities expand, they face issues such as inadequate housing, traffic congestion, pollution, and strained public services. The rapid influx of people often leads to informal settlements with substandard living conditions, limited access to sanitation, clean water, or healthcare facilities, all of which can exacerbate health risks and reduce quality of life [2, 3, 9].

Public health challenges in rapidly urbanizing megacities are compounded by their high population density, which facilitates the spread of infectious diseases. Crowded living conditions, inadequate ventilation, and poor waste management increase the transmission rates of diseases such as tuberculosis, dengue fever, and respiratory infections. In addition, urban areas are often hotspots for non-communicable diseases due to sedentary lifestyles, poor air quality, and limited access to green spaces. The COVID-19 pandemic highlighted how swiftly diseases can spread in densely populated cities, where social distancing is challenging, and healthcare systems can be easily overwhelmed. To create healthier urban environments, there is a pressing need to address these public health challenges through integrated urban planning and policy measures that prioritize the health and well-being of urban populations [1, 6, 8, 31].

1.2 Population Mobility and Public Health in Megacities

Population mobility is a defining feature of modern megacities, driving economic growth, innovation, and cultural exchange. As people move within and between urban areas, they contribute to dynamic economic activities, diverse social interactions, and the overall vitality of city life. However, this constant movement also presents significant challenges, particularly regarding public health. High mobility rates can facilitate the rapid spread of infectious diseases, making containment efforts more complex and requiring tailored public health strategies that address the unique needs of a highly mobile urban population[3, 9, 10, 15, 16].

Managing public health in megacities necessitates a comprehensive understanding of how population mobility intersects with disease transmission dynamics. Traditional health interventions, such as quarantine or vaccination campaigns, are less effective without considering the spatial and temporal patterns of human movement. To ensure healthier urban environments, public health planning must integrate mobility data to identify high-risk areas and prioritize resources effectively. This integration is especially critical in megacities like Shenzhen, where complex transportation networks, including metro systems, significantly influence population mobility patterns and, consequently, the potential for disease spread [3, 9, 15].

1.3 Metro Systems as Conduits for Disease Transmission

Metro systems, as vital arteries of modern urban transportation, play a critical role in shaping population mobility in megacities. While they provide efficient and affordable mobility, their high passenger volumes and confined spaces can also serve as conduits for the rapid spread of infectious diseases. During peak hours, metro systems often facilitate close physical contact among passengers, increasing the likelihood of transmission through respiratory droplets or contact with contaminated surfaces. The frequent turnover of passengers, combined with the extensive reach of metro networks, creates multiple opportunities for pathogens to spread across different city areas [7, 12, 18, 20].

For instance, during the COVID-19 pandemic, metro systems in cities like New York, Tokyo, and London were identified as high-risk environments due to the concentration of passengers in confined spaces for extended periods. This highlighted the potential of metro systems to act as vectors for disease spread, especially in densely populated urban areas. In cities like Shenzhen, where metro systems are heavily relied upon for daily commuting, these networks have become focal points for public health monitoring and intervention. Traditional public health measures, such as social distancing and hygiene protocols, often face challenges in these settings due to space limitations and high passenger volumes. As a result, understanding how metro systems contribute to the dynamics of epidemic transmission is crucial for developing effective disease control strategies [18, 20, 23, 27, 31].

Research has shown that specific factors, such as ventilation quality, crowd density, and passenger interaction patterns, can significantly influence transmission risks within metro environments. Analyzing these factors through detailed metro data can provide urban planners and public health officials with insights into potential hotspots for disease spread, enabling more targeted interventions. By integrating transportation data with epidemiological models, cities can better predict and mitigate the impact of infectious diseases, enhancing their resilience against future health crises [23, 27, 31].

1.4 Novel Approaches Leveraging Metro Data for Healthier City Development

This chapter introduces new ways of understanding the relationship between urban mobility and public health by leveraging metro data. It focuses on how analyzing passenger movement patterns in metro systems, particularly in megacities like Shenzhen, can help identify areas with heightened risks of infectious disease spread. By doing so, it provides practical guidance for city planners and policymakers on designing healthier, safer urban spaces, demonstrating how innovative data usage can contribute to building resilient urban environments [19, 22, 14, 18].

Moreover, the chapter emphasizes the importance of integrating transportation data into public health strategies, fostering collaboration among urban planners, public health officials, and data scientists. This interdisciplinary approach can support the development of comprehensive strategies for managing population mobility and mitigating health risks, ultimately guiding cities to be more adaptable and prepared for future public health challenges [25, 26, 29, 30].

2 Conceptual Framework: Understanding the Interplay Between Mobility and Health

2.1 Population Mobility and Urban Development

Population mobility is a dynamic force shaping urban development and influencing public health outcomes. In rapidly growing megacities, the constant movement of people is both a catalyst for economic activity and a vector for the spread of infectious diseases. Literature on urban sociology, such as the work by [1, 2], emphasizes the role of mobility in fostering innovation, social interaction, and economic growth. However, increased mobility also brings challenges, including congestion, pollution, and heightened exposure to infectious diseases.

Recent studies on the dynamic features of population mobility highlight how the movement patterns of individuals, especially in dense urban settings, can influence the spread of diseases like COVID-19. Mobility data derived from mobile phones, metro card swipes, and GPS tracking have been used to model disease transmission dynamics. This research shows that frequent movement between urban centers and peripheral areas can lead to rapid, widespread outbreaks, especially when public health measures like social distancing or mask-wearing are not adequately enforced. Understanding these mobility patterns is critical for developing effective strategies to manage public health risks in urban environments [9, 16, 11].

Moreover, studies have explored how different types of mobility—such as commuting, leisure travel, or emergency evacuation—affect urban resilience and public health. It is demonstrated that not all mobility is equal; short, repetitive trips within a city may pose a different risk than longer, less frequent intercity travel. Such findings underscore the need for nuanced urban planning that considers the specific types of mobility and their potential impacts on public health [17, 13, 16, 28].

2.2 *Metro Systems and Disease Transmission*

Metro systems serve as both vital transportation networks and potential conduits for disease transmission due to their unique characteristics. Research on the role of public transit in disease spread emphasizes the high-density, high-turnover nature of metro systems, which can accelerate the transmission of airborne diseases like COVID-19, influenza, and tuberculosis. For example, studies have shown that metro lines with higher passenger volumes and longer dwell times at stations tend to have greater risks for disease transmission [21, 23, 27, 31].

The dynamic features of metro-based mobility, captured through smart card data and mobile sensors, have been used to model and predict the spread of infectious diseases in real time. A study by Zhang et al. [3] analyzed metro card data from Beijing to assess the risk of disease transmission across different lines and stations, revealing that certain stations functioned as critical nodes in the network, with higher passenger flows and greater risks of infection. These findings have led to targeted public health interventions, such as increased ventilation, regular disinfection, and passenger flow control in high-risk stations [18, 20, 21, 23].

During the COVID-19 pandemic, cities like New York, Tokyo, and London highlighted the importance of metro systems in epidemic dynamics, where certain lines and stations were identified as high-risk due to high passenger density and prolonged exposure in confined spaces. The literature also discusses the effectiveness of various interventions, such as physical distancing, mask mandates, and real-time monitoring, in reducing transmission risks. Public health researchers suggest that these measures should be integrated into broader urban planning strategies to create safer public transport environments and minimize the public health risks associated with high-density transit networks.

Furthermore, studies suggest that understanding the behavioral patterns of metro users, such as peak travel times, route choices, and response to public health messaging, can help tailor interventions to specific contexts. This research points to the value of an interdisciplinary approach, combining insights from epidemiology, urban planning, and behavioral science, to address the complex dynamics of disease transmission in metro systems [7, 18, 24, 27, 29].

2.3 Innovative Approaches in Public Health and Urban Planning

Innovative approaches in public health and urban planning have emerged in response to the challenges posed by infectious diseases in densely populated cities. These approaches leverage new technologies and interdisciplinary methodologies to better understand and mitigate epidemic risks. Recent literature highlights several promising strategies, such as using big data, artificial intelligence (AI), and digital epidemiology to monitor and predict outbreaks in real time. For example, studies by [4, 5] illustrate how mobility data from smartphones, public transport, and social media can be integrated to create dynamic models that forecast the spread of diseases. Such models help identify potential hotspots and predict future trends, enabling targeted public health interventions.

AI and machine learning algorithms enhance understanding of epidemic dynamics by analyzing vast amounts of health and mobility data to provide real-time insights. Natural language processing techniques can extract valuable signals from social media, detecting early signs of outbreaks, while deep learning models refine epidemiological predictions. These advancements offer new ways to optimize resource allocation and develop proactive health measures during crises.

Digital epidemiology, which leverages digital tools to understand disease dynamics, provides urban planners with critical insights into contact patterns, travel behavior, and social interactions. This approach has been instrumental during the COVID-19 pandemic, allowing for more precise interventions based on localized disease spread patterns.

Metro data, as a component of urban mobility data, plays a particularly significant role in these innovative approaches. The dense passenger traffic and interconnected nature of metro systems make them both a facilitator of rapid disease transmission and a rich source of data. By analyzing metro data, including passenger flows, travel frequencies, and station occupancy rates, public health officials can identify high-risk areas and times, enabling more targeted interventions. This approach was utilized during COVID-19 in cities like Seoul and Tokyo, where real-time metro data helped manage crowding and direct sanitation efforts more effectively.

Moreover, technological advancements enable the creation of "smart health cities," where integrated urban data systems, including transportation networks, contribute to city resilience. In cities like Singapore, combining metro data with health monitoring tools and digital contact tracing has proven effective in controlling outbreaks by swiftly identifying and isolating cases, underscoring the value of transportation data in public health strategy.

Metro data analysis involves several key methods: passenger flow monitoring through automated fare collection data, travel pattern identification using smart card data, and real-time surveillance using sensors. Advanced statistical tools and spatial modeling techniques can further analyze this data, pinpointing high-risk zones and predicting the likelihood of transmission based on crowd density and movement patterns.

This review of existing and emerging methods, including the strategic use of metro data, underscores the importance of integrating diverse data sources and technologies in shaping healthier, more resilient urban environments. These insights provide a foundation for understanding how data-driven approaches can mitigate epidemic risks and enhance public health in megacities, setting the stage for a deeper exploration of metro data's role in epidemic management in the following section.

3 Methodology: Analyzing Metro Data for Epidemic Risk Reduction

3.1 Case Study: Shenzhen City

Shenzhen City, a rapidly expanding megacity in southern China (see Fig. 1), offers a unique context for studying the interplay between population mobility and epidemic risk due to its dynamic growth, high density, and extensive metro network. With a population exceeding 17 million, Shenzhen represents a complex urban environment where the movement of people, goods, and services is facilitated by an expansive and heavily used metro system. This public transit infrastructure, which includes 16 metro lines and over 300 stations, serves as the lifeline for daily commutes, connecting various residential, commercial, and industrial zones.

The city's rapid urbanization has brought both opportunities and challenges. While economic growth has surged, so has the complexity of managing public health risks in a densely populated area. Shenzhen's experience during the COVID-19 pandemic, where several outbreaks were traced back to metro-related clusters, underscores the need to better understand how metro systems can both facilitate mobility and pose risks for infectious disease spread. The metro system's role in transporting millions

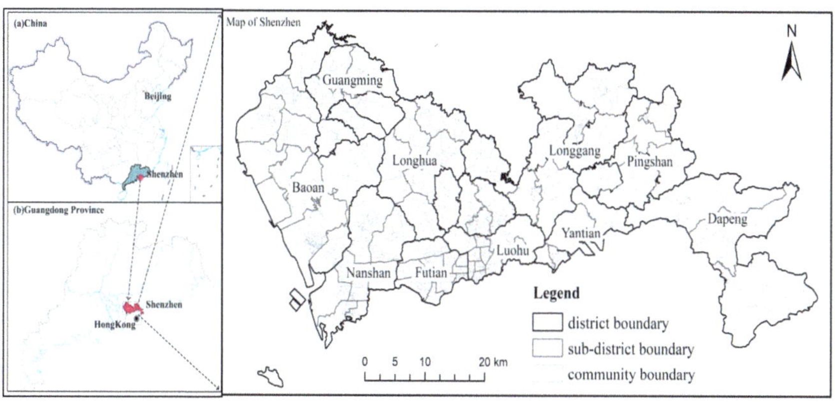

Fig. 1 Map of Shenzhen city as the study

of passengers daily makes it a critical focal point for analyzing disease transmission patterns, particularly in high-density urban environments.

This chapter utilizes Shenzhen as a case study to explore how metro data can inform strategies for healthier city development. The case study approach allows for a deep dive into real-world dynamics, providing insights that can be generalized to other megacities facing similar challenges. By focusing on Shenzhen, the chapter draws from a rich dataset and a variety of methodological approaches to analyze population mobility, identifying specific risks and opportunities for public health interventions in urban transit networks.

3.2 *Data Collection*

To comprehensively analyze the relationship between metro use and epidemic risk, the study employs a multi-layered data collection strategy, integrating various sources of metro and mobility data. The primary dataset comprises smart card transaction records from Shenzhen Metro, detailing passenger entries, exits, travel routes, and transit durations. This granular data captures daily mobility patterns, offering a rich source of information on passenger density and flow within the metro system (see Fig. 2).

In addition to transaction data, the study leverages GPS-based data from mobile applications, which provides real-time insights into passenger movement across different transport modes, including transitions between metro use and other forms of transportation. This integration of multiple data sources allows for a more comprehensive view of how people move through the city and interact with the metro network, enhancing the ability to model and predict transmission risks.

Demographic data, including population density, age distribution, and socioeconomic indicators at the community level, are also incorporated to better understand the underlying factors influencing disease spread. Public health data, such as

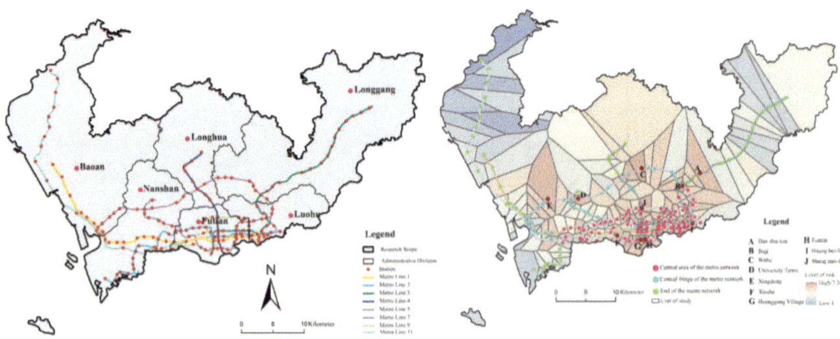

Fig. 2 Map of Shenzhen metro system (l) and division of metro network structure (r)

COVID-19 case reports, is cross-referenced with mobility data to identify correlations between movement patterns and outbreak clusters. Moreover, spatial data, including the geographic locations of metro stations, is used to map and visualize the spatial distribution of risks and identify "hotspots" where interventions may be most needed.

To ensure data accuracy, validation is conducted through real-time monitoring tools, such as sensors, and crowd-sourcing platforms that report on passenger volumes and behaviors at metro stations. This data triangulation helps confirm the reliability of the findings and supports robust analysis.

By combining these diverse datasets, the study is able to provide a detailed and dynamic picture of population mobility and its impact on public health in Shenzhen. The insights gained from this comprehensive data collection process are critical for developing targeted strategies that can mitigate epidemic risks and enhance urban resilience.

3.3 Analytical Framework

The analytical framework combines statistical, spatial, and simulation methods to examine the relationship between metro mobility patterns and epidemic risk. Statistical techniques, such as regression analysis and time-series modeling, are employed to quantify the impact of metro use on infection rates while controlling for various contextual factors, including demographic attributes, time variations, and station characteristics (Fig. 3).

Spatial analysis, using Geographic Information System (GIS) tools, maps the geographic distribution of COVID-19 cases in relation to metro stations and lines. This enables the identification of "hotspots" where disease transmission is more likely due to high passenger volumes or specific network characteristics. Techniques

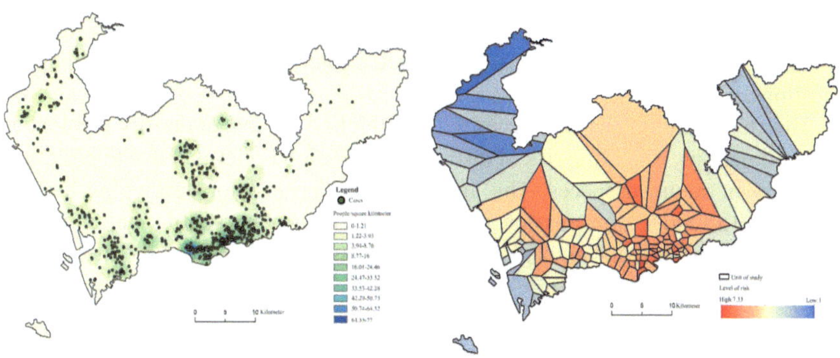

Fig. 3 Spatial distribution of local COVID-19 cases (l) and epidemic risk levels (r) in Shenzhen

like spatial clustering and spatial autocorrelation are utilized to uncover underlying patterns and dynamics.

In addition to these methods, the framework incorporates predictive modeling techniques to evaluate the effectiveness of different intervention strategies. These models simulate various outbreak scenarios, taking into account changes in mobility patterns, public health interventions, and policy measures like mask mandates and social distancing.

This integrated approach allows for a nuanced understanding of how metro systems contribute to epidemic dynamics and offers actionable insights for urban planning and public health strategies. The findings provide a foundation for developing targeted interventions that enhance urban resilience and reduce public health risks in densely populated megacities like Shenzhen.

4 Findings and Discussion: Insights from Metro Data in Shenzhen

4.1 Patterns of Population Mobility in Shenzhen

Analyzing the metro data from Shenzhen reveals several key patterns of population mobility that are crucial for understanding epidemic risks. The data shows a concentration of passenger traffic during peak hours, with a significant number of trips originating from and ending at high-density residential and commercial areas. Key metro stations, particularly those located in central business districts and major residential zones, exhibit the highest passenger volumes, creating potential hotspots for disease transmission.

The analysis also highlights distinct mobility patterns during weekdays and weekends, with weekday travel dominated by work-related commutes, while weekends show more dispersed movements for leisure and shopping activities. These patterns provide insights into how different types of mobility behaviors contribute to the spread of infectious diseases, enabling targeted interventions at specific times and locations.

The study further identifies areas with the highest passenger density and longest transit durations, where the risk of disease transmission is elevated due to prolonged exposure in confined spaces. This information is critical for developing targeted strategies to reduce transmission risks, such as improving ventilation, enhancing cleaning protocols, and managing crowd flow through real-time monitoring and dynamic routing adjustments.

4.2 Impact of Population Mobility on Epidemic Transmission

The analysis of metro data reveals a strong correlation between mobility patterns and the transmission of infectious diseases in Shenzhen. Metro lines and stations that serve densely populated residential and commercial areas are associated with higher transmission risks. Stations with the highest passenger throughput were found to overlap with areas that reported higher numbers of COVID-19 cases, indicating that metro use plays a significant role in the spatial distribution of outbreaks.

Particularly, stations with high transfer rates—where passengers frequently switch lines or wait for connecting trains—emerge as critical points of concern. These stations not only experience higher passenger density but also longer dwell times, increasing the likelihood of close physical contact and exposure. The findings suggest that mobility patterns within the metro system directly influence the spread of diseases, especially in high-density environments.

The data analysis also highlights temporal patterns, with higher risks observed during peak commuting hours when passenger volume is at its highest. These findings offer crucial insights into the dynamics of epidemic transmission, underscoring the importance of understanding both the spatial and temporal aspects of population mobility in urban settings.

While these findings are significant in identifying "hotspots" and times of elevated risk, they also emphasize the need for ongoing monitoring and adaptive strategies to mitigate transmission. The detailed insights from the metro data provide an evidence base for targeted public health interventions, focusing on specific high-risk stations and times.

5 Policy Implications: Integrating Transportation and Health Planning

5.1 Strategies for Healthier City Development

Based on metro data analysis in Shenzhen, several targeted interventions emerge that can significantly reduce epidemic risks while fostering healthier city development. Short-term interventions include upgrading metro system ventilation using HEPA filters and UV disinfection to improve air quality and reduce airborne transmission risks. Metro operators should also implement enhanced hygiene protocols, with regular disinfection of high-contact surfaces, to minimize the risk of contamination.

Crowd management strategies are critical in minimizing the impact of peak-hour congestion. By leveraging real-time passenger flow data, trains can be scheduled more frequently, and smart-routing technologies can help distribute passengers more evenly across stations. Designing waiting areas to ensure social distancing during peak times and developing infrastructure for controlled entry and exit at metro stations can further help reduce the risk of close contact between passengers.

In terms of long-term urban planning strategies, city planners should prioritize integrating metro systems with active transportation options like cycling and walking. Encouraging short-distance trips on foot or by bike can reduce reliance on the metro for short journeys, alleviating pressure on the system. Infrastructure enhancements, such as expanding bike-sharing programs and providing safe pedestrian pathways, can help achieve this shift while promoting healthier lifestyles.

Urban planning should also consider designing metro stations as health-focused urban nodes. Metro stations could serve dual roles as both transport hubs and health information centers, where commuters have access to public health messaging, vaccination drives, or basic health screenings. This integration of health services into daily commutes makes public health interventions more accessible and convenient for the urban population.

5.2 Recommendations for Policymakers

The findings underscore the importance of real-time data integration between metro systems and public health agencies. Such data exchange enables rapid responses to emerging health threats by tracking mobility patterns and identifying high-risk zones. Developing robust frameworks for real-time communication and data sharing can help public health officials implement targeted interventions more efficiently, whether it's for localized containment or broader public health messaging.

Policymakers should also explore the potential of digital tools, such as mobile health applications that provide real-time alerts on crowding or infection risk levels at specific metro stations. These tools empower commuters to make informed decisions about their travel plans, reducing unnecessary exposure. Additionally, investments in digital infrastructure for monitoring and responding to passenger flow are essential to support automated alerts for overcrowding and facilitate more responsive transit operations.

Furthermore, there is a need for public health guidelines that regulate metro hygiene and passenger safety. Establishing strict disinfection protocols, mandating the use of personal protective equipment (PPE) during outbreaks, and promoting public awareness campaigns will bolster the effectiveness of other interventions. Public health agencies should collaborate with metro operators to implement these guidelines seamlessly across all transit networks.

5.3 Collaboration and Implementation

Effective implementation of these strategies hinges on collaboration between urban planners, public health experts, and transportation authorities. Policymakers must foster partnerships that include government agencies, private sector stakeholders, research institutions, and community organizations. These interdisciplinary working groups can co-develop strategies that address the dual challenges of urban mobility and public health.

For example, public–private partnerships could facilitate the development of smart technologies like automated crowd monitoring systems, while partnerships with academic institutions could advance the predictive models needed to forecast epidemic outbreaks based on mobility data. Collaborative frameworks should also be established to guide long-term urban planning decisions that prioritize health and resilience.

Public engagement plays a crucial role in successful implementation. Raising awareness among commuters about safe travel practices through education campaigns, signage, and digital platforms ensures public cooperation and compliance with health guidelines. Creating feedback mechanisms where community members can voice concerns or suggest improvements will allow policymakers to adjust strategies as needed, keeping interventions relevant and effective.

Funding mechanisms are also vital. Local governments should explore opportunities for financial support from national health programs, international grants, or public–private partnerships dedicated to sustainable urban development. Regular monitoring, evaluation, and adaptive management are necessary to refine these strategies over time and ensure they remain effective in mitigating risks.

By fostering a multi-sectoral approach and aligning public health and transportation planning, cities can leverage resources and expertise to build more resilient urban infrastructures that can withstand future epidemics. Long-term collaboration will help ensure that the lessons learned from this analysis are translated into sustainable urban health practices.

6 Conclusion: Building Resilient and Healthy Urban Futures

This chapter has illustrated the pivotal role of metro data in shaping healthier and more resilient cities. By analyzing patterns of population mobility, we have demonstrated that metro systems, while essential for urban connectivity, also act as key conduits for disease transmission, especially in megacities like Shenzhen. The insights gained from metro data enable targeted interventions, such as enhanced ventilation, strategic crowd management, and real-time data-driven measures, all of which contribute to reducing epidemic risks.

Integrating transportation and public health planning is crucial for cities to not only respond effectively to health emergencies but also foster long-term urban resilience. This chapter underscores the need for interdisciplinary approaches that incorporate public health perspectives into urban transit planning, a strategy that can significantly mitigate public health risks in densely populated environments.

Future research should explore the application of advanced technologies, such as artificial intelligence and machine learning, to refine predictive models for epidemic outbreaks. Expanding this research to other megacities with different urban forms and transit systems will provide broader insights into how mobility data can inform urban planning and public health strategies. Furthermore, the long-term health impacts of integrating public health goals into transit-oriented development should be closely studied, with a focus on sustainable urban design principles.

The COVID-19 pandemic has served as a reminder that urban resilience is deeply connected to how we manage mobility in our cities. By leveraging innovative approaches, such as metro data analysis, we can build cities that are not only more efficient but also better prepared to face future health challenges. The interdisciplinary strategies outlined in this chapter offer valuable guidance for policymakers, urban planners, and public health professionals alike, providing a roadmap toward creating urban environments that prioritize the health and well-being of all citizens.

References

1. Castells M (2000) The rise of the network society: the information age: economy, society, and culture, 2nd edn. Wiley-Blackwell, London
2. Glaeser EL (2011) Triumph of the city: how our greatest invention makes us richer, smarter, greener, healthier, and happier. Penguin Press, New York
3. Zhang J, Cheng Y, Wang L, Li X (2020) Urban resilience and public health planning in the face of pandemics. J Urban Health 97(2):1
4. Aiello AE, Renson A, Zivich PN (2021) Social media- and internet-based disease surveillance for public health. Annu Rev Public Health 42:361–375
5. Chinazzi M, Davis JT, Ajelli M, Gioannini C, Litvinova M, Merler S, Pastore y Piontti A, Mu K, Rossi L, Sun K, Viboud C, Xiong X, Yu H, Halloran ME, Longini IM, Vespignani A (2020) The effect of travel restrictions on the spread of the 2019 novel coronavirus (COVID-19) outbreak. Science 368(6489):395–400
6. Allam Z, Jones DS (2020) Pandemic stricken cities on lockdown: Where are our planning and design professionals now, then, and into the future? Land Use Policy 97:1
7. Behram W, Frank LD (2021) Neighborhood-level COVID-19 hospitalizations and mortality relationships with built environment, active, and sedentary travel. Health Place 71:1
8. Bo L, Liu Q, He H, Peng Y (2021) Built environment and early infection of COVID-19 in urban districts: a case study of Huangzhou. Sustain Cities Soc 66:1
9. Campello M, Kankanhalli G, Muthukrishnan P (2023) Corporate hiring under COVID-19: financial constraints and the nature of new jobs. J Financ Quant Anal 58(4):1
10. Choi KH, Denice P (2022) Socioeconomic variation in the relationship between neighborhoods' built environments and the spread of COVID-19 in Toronto, Canada. Can Stud Popul 49:1
11. Credit K (2020) Neighbourhood inequity: exploring the factors underlying racial and ethnic disparities in COVID-19 testing and infection rates using ZIP code data in Chicago and New York. Reg Sci Policy Pract 12(6):1
12. Dietz L, Horve PF, Coil DA, Fretz M, Eisen JA, Van Den Wymelenberg K (2020) 2019 novel coronavirus (COVID-19) pandemic: built environment considerations to reduce transmission. mSystems 5(2):1
13. DiMaggio C, Klein M, Berry C, Frangos S (2020) Black/African American communities are at highest risk of COVID-19: spatial modeling of New York City ZIP code-level testing results. Ann Epidemiol 51:1
14. Duan J, Yang B, Zhou L, Zhang J (2020) Planning improves city's immunity: a written conversation on COVID-19 breakout. City Plan Rev 44(6):1

15. Emre T (2023) The impact of built and socio-economic environment factors on COVID-19 transmission at the ZIP-Code level in Florida. J Environ Manag 326:1
16. Gaisie E, Oppong-Yeboah NY, Cobbina PB (2022) Geographies of infections: built environment and COVID-19 pandemic in metropolitan Melbourne. Sustain Cities Soc 81:1
17. Guo Y, Yu H, Zhang G, Ma DT (2021) Impacts of travel-implied policy factors on COVID-19 spread within communities based on multi-source data interpretations. Health Place 69:1
18. Hu Y, Lin Z, Jiao S, Zhang R (2024) Urban form and COVID-19 transmission dynamics: assessing the impact of metro station accessibility and passenger flow in high-density cities. Buildings 14(1):103
19. Jin D, Zhang J, Cheng Y, Zhou G, Liu Z (2022) What lesson does COVID-19 teach planning? Urban Plan Forum 23(2):1
20. Kan Z, Kwan MP, Wong MS, Huang J, Liu D (2021) Identifying the space-time patterns of COVID-19 risk and their associations with different built environment features in Hong Kong. Sci Total Environ 772:1
21. Liu C, Guan C, Liu Z (2021) The impacts of the built environment on the incidence rate of COVID-19: a case study of King County. Sustainable Cities and Society, Washington, DC, p 74
22. Navaratnam S, Nguyen K, Selvaranjan K, Zhang G, Mendis P, Aye L (2022) Designing post-COVID-19 buildings: approaches for achieving healthy buildings. Buildings 12(1):1
23. Niu Q, Wu W, Shen J, Huang J, Zhou Q (2021) Relationship between built environment and COVID-19 dispersal based on age stratification: a case study of Wuhan. Int J Environ Res Public Health 18(14):1
24. Pinter-Wollman N, Jelić A, Wells NM (2018) The impact of the built environment on health behaviours and disease transmission in social systems. Philos Trans R Soc B Biol Sci 373(1753):1
25. Quynh CN, Huang Y, Kumar A, Duan H, Jay J (2020) Using 164 million Google Street View images to derive built environment predictors of COVID-19 cases. Int J Environ Res Public Health 17(17):1
26. Rahman MH, Zafri NM, Ashik FR, Khan A (2021) Identification of risk factors contributing to COVID-19 incidence rates in Bangladesh: a GIS-based spatial modeling approach. Heliyon 7(4):1
27. Schmiege D, Haselhoff T, Ahmed S, Anastasiou OE, Moebus S (2023) Associations between built environment factors and SARS-CoV-2 infections at the neighbourhood level in a metropolitan area in Germany. J Urban Health 100(1):1
28. Sokadjo YM, Atchadé MN (2020) The influence of passenger air traffic on the spread of COVID-19 in the world. Transp Res Interdiscip Perspect 8:1
29. Tribby CP, Hartmann C (2021) COVID-19 cases and the built environment: initial evidence from New York City. Profess Geographer 73(3):1
30. Wang J, Zeng F, Tang H, Wang J, Xing L (2022) Correlations between urban built environmental factors and the spatial distribution at the community level in the reported COVID-19 samples: a case study of Wuhan. Cities 129:1
31. Zheng T, Liu H (2020) Exploration of the built-environmental elements that influence the spread of COVID-19 pandemic on community scale: a case study of Wuhan, China. Modern Urban Res 20:1

Strengthening Health-Resilience Nexus Through Sustainable Urban Design in Brisbane, Australia

Ali Cheshmehzangi, Baojie He, and Tong Zuo

Abstract This final chapter explores the crucial connection between resilience and health in the framework of urban design, with a particular emphasis on Brisbane, Australia. It highlights how crucial it is to incorporate health concerns into urban planning in order to improve public health outcomes and fortify community resilience in the face of issues like socioeconomic inequality and climate change. The chapter demonstrates how sustainable urban design can encourage social cohesion, active living, and improved food security through case studies, such as the Norman Creek flood mitigation project in Brisbane and the Kelvin Grove Urban Village. Important discoveries show that the relationship between health and resilience is greatly impacted by urban planning, emphasizing the necessity for legislative frameworks that give inclusive, accessible, and ecologically friendly designs top priority. In order to improve health resilience in urban planning, the chapter ends with policy recommendations. It also makes recommendations for future research directions to better understand the relationship between health and urban resilience. In the end, it promotes an integrated approach to urban planning that puts everyone's health first and takes long-term sustainability objectives into account.

Keywords Urban design · Health resilience nexus · Sustainable urban design · Brisbane · Climate change · Flooding · Healthy cities · Spatial

---

A. Cheshmehzangi (✉) · B. He
School of Architecture, Design and Planning, The University of Queensland, Brisbane, Australia
e-mail: A.Chesh@uq.edu.au

B. He
Centre for Climate-Resilient and Low-Carbon Cities, Chongqing University, Chongqing, China

T. Zuo
Department of Agro-Food Sciences and Technologies (DISTAL), University of Bologna Alma Mater Studiorum, Viale Fanin, Bologna, Italy

A. Cheshmehzangi et al. (eds.), *Designing Healthy Cities*, Urban Sustainability,
https://doi.org/10.1007/978-981-96-5538-0_10

1 Introduction: An Overview of the Health-Resilience Nexus in Urban Design

One essential response to the increasing complexity of contemporary cities is the integration of health and resilience into (sustainable) urban design [1, 2]. The relationship between public health and urban resilience—where sustainable urban design not only improves the general well-being of communities but also lessens the effects of social and environmental stressors—is known as the "health-resilience nexus" [3–8]. In an increasingly urbanising world, where cities must contend with issues like natural disasters, climate change, and socioeconomic inequality, it is imperative that health-focused resilience be ingrained in urban planning [9, 10]. This closing chapter examines this relationship in the context of Brisbane, Australia, a city under increasing pressure from spatial inequality, fast population growth, and climate risks.

The relationship between health and resilience in urban planning pertains to how resilient systems can be fostered in urban environments to improve public health outcomes [11, 12]. Cities are said to be resilient if they can absorb, bounce back from, and adjust to different kinds of disturbances, whether they be social, economic, or environmental, without endangering the welfare of their citizens. In contrast, the World Health Organization (WHO) defines health as the state of being physically, mentally, and socially well in addition to being free from disease [13]. This nexus shows that improving health outcomes, particularly for vulnerable populations, is closely correlated with the resilience of urban infrastructure, governance, and social systems [14, 15]. For instance, the layout of green areas, pedestrian-friendly streets, and interconnected neighbourhoods encourage social cohesion, active living, and mental health in addition to lowering the risk of flooding and the urban heat island effect [16]. Likewise, public health risks like heat stress, respiratory disorders, and vector-borne infections are decreased by strategies that address resilience to climate events like heatwaves and flooding [17, 18]. The correlation between resilience and health highlights the significance of adopting comprehensive approaches to urban design that tackle both short-term health issues and long-term urban sustainability.

1.1 Relevance of the Health-Resilience Nexus in Sustainable Urban Planning

By definition, sustainable urban planning seeks to develop cities that are both liveable now and capable of overcoming obstacles in the future. It is becoming more and more obvious that Brisbane needs an urban environment that is resilient to health issues [19–21]. According to climate change projections, Brisbane will see an increase in the frequency and severity of extreme weather events, such as heatwaves, floods, and storms [22]. Public health is at risk from these occurrences both directly and indirectly [23], from heat-related illnesses to the spread of waterborne infections during floods.

Brisbane, like many other cities, is also dealing with socio-spatial disparities that have an impact on health resilience [24]. Health disparities exist between different regions because outer suburbs frequently lack access to green spaces, healthcare facilities, and public transportation [25]. By approaching these issues from the perspective of health resilience, urban areas can be designed to be resilient to shocks to the environment while simultaneously actively enhancing the health of all inhabitants. These gaps can be filled in part by implementing sustainable urban design strategies that incorporate equitable access to amenities, water-sensitive urban design (WSUD), and nature-based solutions (NBS).

1.2 Brisbane's Urban Design Initiatives and the Health-Resilience Nexus

Significant progress has already been made by Brisbane to incorporate resilience and health into its framework for urban planning [26]. The city's Resilient Brisbane strategy places a strong emphasis on the necessity of safeguarding residents' social well-being while planning for potential future climate effects [27, 28]. Brisbane is using sustainable urban design to improve resilience and public health through initiatives like the Norman Creek flood mitigation initiative and the park's transformation [29]. Along with lowering the risk of flooding and reducing heat, these projects also provide community spaces that promote social interaction and physical activity, all of which lead to healthier lifestyles (Figs. 1, 2, 3).

One excellent example of the relationship between health and resilience in action is Brisbane's emphasis on using natural solutions as part of its efforts to increase resilience [30]. The city is addressing environmental risks and enhancing public health at the same time by investing in wetlands, urban forests, and other green infrastructure projects [31]. For example, protecting and growing urban green spaces are important ways to reduce heat islands, which can cause illnesses associated with the heat. Additionally, these areas are crucial locations for leisure pursuits that improve residents' mental health and strengthen their bonds with one another.

Furthermore, Brisbane's CityShape 2026 project demonstrates the city's dedication to promoting inclusive, walkable, and connected communities [32]. Reducing emissions and improving air quality—two important components of resilience—while also encouraging healthier lifestyles through the city's prioritisation of active transportation infrastructure, such as bike lanes and pedestrian-friendly streets [33]. These initiatives demonstrate the city's understanding that urban planning significantly affects public health and resilience to environmental stressors.

Figs. 1, 2 and 3 Example of UQ lakes at the University of Queensland, a combination of green space, functional green and blue infrastructure, and healthy living spaces (Author's Own)

1.3 Objective and Scope of the Chapter

The objective of this closing chapter is to present a thorough analysis of how Brisbane's urban design can improve the relationship between health and resilience. It demonstrates the ways in which urban infrastructure, green spaces, active transportation, and community-level interventions (such as urban farming) support resilience and public health by examining significant spatial case studies and fact-based examples. This chapter examines particular projects in Brisbane that show how health and resilience can be integrated, offering insights into how these approaches can be expanded or modified for use in different urban contexts [34]. In the end, it aims to motivate researchers, policymakers, and urban planners to give health-resilience connections top priority when creating sustainable, future-ready cities. The chapter also serves as the closing chapter of the book, highlighting the importance (and relevance) of health-resilience nexus in urbanism, particularly through sustainable urban design.

2 Urban Health and Resilience Challenges in Brisbane

2.1 Climate Vulnerabilities and Health Impacts

Brisbane, which has a subtropical climate, is more susceptible to the effects of climate change, which makes the city's current problems with the environment and public health worse. With an increasing population and urban density, Brisbane is more vulnerable to extreme weather events like storms, heat waves, and floods [35]. These climate vulnerabilities seriously jeopardise public health in addition to upsetting the urban fabric. Given that climate change can have significant negative effects on health, especially for vulnerable populations, addressing these interrelated issues requires an approach to urban planning that prioritises health resilience.

Brisbane is particularly vulnerable to heatwaves and flooding due to its geographic location and climate. The Brisbane River, which flows through the center of the city, frequently overflows during periods of intense rainfall; this problem is made worse by the fact that climate change is increasing the intensity of rainfall [36–38]. There have been several notable floods in the city, including those that occurred in 1974, 2011, and 2022, all of which caused extensive damage and affected thousands of people. The frequency and intensity of these floods are predicted to rise with climate change, putting the city's public health, infrastructure, and economy at serious risk.

Brisbane regularly experiences heatwaves, when summertime temperatures rise above 35 °C. This is another major problem. According to climate projections, heatwaves will get longer and stronger, which will have a greater impact on the elderly, kids, and people with pre-existing medical conditions. In densely populated neighbourhoods, the risk of heat stress is further increased by Brisbane's urban heat island effect, which traps heat in built-up areas.

2.2 *Public Health Impacts of Climate Change*

Brisbane is experiencing a wide range of health effects from climate change, affecting both mental and physical health [23]. Heat stress is a serious issue, especially for vulnerable groups like the elderly, those suffering from long-term illnesses, and those residing in low-income areas with little access to cooling facilities. Dehydration, heat exhaustion, and even heatstroke can result from heatwaves and can be fatal if left untreated. Heat waves also exacerbate poor air quality, especially in places with heavy traffic, which makes respiratory diseases like asthma and other long-term lung disorders worse.

Flooding presents hazards to the public's health in the short and long term. Floods increase the risk of getting hurt, drowning, and coming into contact with tainted water, which can result in the spread of diseases like gastroenteritis and leptospirosis [39, 40]. Following floods, standing water can also serve as a mosquito breeding ground, raising the possibility of contracting diseases like dengue fever that are spread by vectors. Furthermore, floods have the potential to impair healthcare services and worsen already-existing health disparities, especially for low-income communities residing in flood-prone areas [41]. Events related to climate change also have significant effects on mental health. In addition to the drawn-out healing process, the trauma of losing loved ones, homes, and belongings during floods can result in long-term psychological stress, anxiety, and depression. These psychological difficulties frequently linger long after the physical effects of climate catastrophes have been mitigated.

2.2.1 Socioeconomic Health Disparities

Like many large cities, Brisbane has notable socioeconomic health disparities in its suburbs, which have an immediate impact on the resilience and general well-being of the community. These differences are directly related to factors like housing, work, education, income, and access to basic services. While those living in Brisbane's outer, lower-income suburbs face more severe health issues and restricted access to care, those living in wealthier, inner-city suburbs typically enjoy better health outcomes and greater access to health services. Regardless of socioeconomic background, resolving these disparities is essential to creating an urban environment that is resilient to health and benefits all inhabitants.

Brisbane's urban landscape spatially reflects the city's socioeconomic divide. Inner-city suburbs like New Farm, West End, and South Brisbane are distinguished by greater access to healthcare and other services, better educational attainment, and higher income levels [42]. These neighbourhoods typically have more green space, are more walkable, and are located closer to medical centers and hospitals. These suburbs offer better access to wholesome food options, better air quality, and a higher likelihood of regular physical activity among their residents. Reduced rates of chronic illnesses like diabetes, heart disease, and respiratory disorders are a result of these factors.

In contrast, there are notable socioeconomic disadvantages in the outer suburbs like Logan, Ipswich, and Inala. Lower income levels, greater unemployment rates, and lower levels of educational attainment are frequently characteristics of these areas [43]. They also typically have less access to specialised medical care, fewer healthcare facilities, and longer travel times to hospitals. Lower rates of private health insurance and a greater reliance on public health systems—which are frequently overburdened in these areas—combine to create a dearth of easily accessible health services.

People who live in these outer suburbs are more likely to be obese, have heart disease, or struggle with mental health issues [44]. The aforementioned conditions are made worse by the restricted availability of preventive healthcare services, like routine examinations, and a deficiency of public facilities that encourage healthy living, like parks and leisure areas. Further exacerbating health risks are the higher levels of environmental degradation and increased exposure to urban heat islands found in many outer suburbs.

2.2.2 Evidence of Health Disparities

Several studies have brought attention to the widening disparities in health that exist in Brisbane's suburbs. According to the 2018 Queensland Health Report, people who live in outer suburbs—especially those in the southwest corridor—have a higher risk of developing chronic illnesses and having worse health outcomes than people who live in inner cities (Queensland [45, 46]). According to the research, differences in access to healthcare, income, and education have been linked to life expectancy differences in certain outer suburbs that are several years lower than in more affluent inner suburbs

Significant disparities in healthcare access throughout Brisbane were also noted in a 2021 Australian Institute of Health and Welfare (AIHW) report [47]. According to the report, people who live in lower socioeconomic suburbs are more likely to rely on bulk-billing general practitioners, who might not be readily available in outlying areas, and to endure longer wait times for public hospital services. The report also highlighted transportation issues, with many people living in the outer suburbs lacking access to sufficient public transportation, which makes it challenging to get to major hospitals or the city center for medical care. Lower-class suburbs have greater rates of mental health problems in addition to physical health inequalities. According to a Brisbane City Council report on mental health from 2022 [48], people who live in the outer suburbs have higher rates of stress, anxiety, and depression. These conditions are frequently brought on by unstable housing, unstable finances, and restricted access to mental health support services.

2.3 Case Study: 2022 Brisbane Floods and Health-Related Aftermath

The floods that hit Brisbane in 2022 are a clear illustration of how vulnerable the city is to health risks brought on by climate change. The Brisbane River burst its banks in February 2022 as a result of intense rains, causing widespread flooding throughout the city. More than 10,000 homes were left uninhabitable, and more than 15,000 properties were submerged [49]. Both short-term and long-term public health effects of the floods were substantial. Hospitals and emergency services were inundated with cases of people exposed to floodwaters, injuries, and drownings in the immediate aftermath. The contaminated drinking water and exposure to floodwaters caused many residents to suffer from skin infections, diarrheal diseases, and other waterborne illnesses. Numerous medical facilities were forced to close or operate at reduced capacity, severely affecting access to healthcare and adding to the already overburdened health services.

The floods had a major negative impact on mental health in addition to these physical health issues. In the months that followed the disaster, a large number of residents reported having high levels of anxiety, depression, and post-traumatic stress disorder (PTSD). The affected residents were assisted by community organisations and mental health services; however, the demand for mental health services far outstripped the supply. Socioeconomic disparities were further aggravated by the floods, as low-income families in flood-prone areas found it difficult to rebuild their lives and recover financially. The floods of 2022 made it abundantly clear that Brisbane must quickly implement a health-resilient urban planning strategy [50, 51], especially in areas that are vulnerable to flooding. Developing a more resilient, health-conscious city requires improvements to flood mitigation strategies, equitable access to healthcare, and urban infrastructure.

3 Sustainable Urban Design for Health Resilience: Global and Local Insights

3.1 Frameworks for Health-Resilient Urban Design

In a time of increasing urbanisation, climate change, and health disparities, city planning is essential to protecting the environment and urban residents' well-being. The goal of health-resilient urban design is to create spaces that actively support people's physical and mental well-being while shielding them from environmental stressors like flooding and intense heat. This section looks at international frameworks that serve as inspiration for these kinds of designs, such as the WHO's recommendations for healthy cities and pertinent Australian laws [52]. It focuses on Brisbane's "Healthy

Cities" initiative, which has adopted creative resilience tactics that are specific to the region.

Cities across the world are realising how crucial it is to incorporate resilience and health into their urban planning. Launched in 1986, the WHO's Healthy Cities initiative is one of the most influential frameworks [53]. This initiative highlights the idea that cities should actively promote well-being by enacting laws that guarantee access to safe, walkable neighbourhoods, clean air, healthcare, and green spaces. Cities should not only prevent disease. The WHO defines health-resilient cities as those that offer fair access to basic services, a safe and clean physical environment, and social cohesion—all of which have a direct bearing on the physical and mental well-being of their citizens [54].

The emphasis on inclusive design—making sure that urban areas meet the needs of everyone, especially vulnerable groups like the elderly, children, and people with disabilities—is a crucial component of the WHO's guidelines [54]. This means that in their design and planning, cities must give priority to accessible green spaces, affordable housing, and effective public transportation. Given that the social fabric of communities is a fundamental component of resilience, the WHO's emphasis on community engagement emphasises [55], even more, how important it is for resilient urban health systems to include citizens in the decision-making process.

UN-Habitat's New Urban Agenda, adopted in 2016, provides guidelines for inclusive and sustainable urban development, and is another significant international project [56–58]. The New Urban Agenda emphasises the inextricable link between public health and urban resilience, advocating for mixed-use, compact, walkable cities [59]. It asks for cutting pollution, improving access to clean water and sanitation, and planning cities that encourage mental health, physical exercise, and environmental sustainability.

3.2 Australian Policies for Health-Resilient Urban Design

Australia has made progress in incorporating resilience and health into its urban planning methods. The Australian Urban Design Protocol, which provides a national guide for creating liveable, healthy, and sustainable cities, is a crucial policy framework driving these initiatives [60]. The guidelines in the protocol prioritise green infrastructure, encourage active transportation, and lower urban heat islands in order to increase resilience and health [61]. In addition to addressing environmental issues, it emphasises the need for accessible, secure, and healthy urban environments.

The National Strategy for Disaster Resilience is another crucial document that highlights how health considerations should be incorporated into resilience planning and climate adaptation [62]. This approach supports the overarching objectives of mitigating disaster risk, ensuring that urban infrastructure is resilient to climate change, and enhancing public health by equipping communities to deal with extreme weather. Locally, specific guidelines for how urban design can support health resilience [63] are provided by state-level initiatives like Queensland's Climate Adaptation Strategy [64]. This approach promotes the integration of natural solutions, like parks and urban wetlands, to improve public health and environmental resilience. These frameworks provide guidance for local planning in Brisbane and serve as a model for incorporating health-resilience concepts into projects involving urban development.

3.3 Case Study: Healthy Cities Brisbane Initiative

Through its Healthy Cities Brisbane initiative, Brisbane has actively embraced the principles of health-resilient urban design. This initiative is part of a larger movement that is in line with the WHO's Healthy Cities framework [54]. The goal of this project is to build a city that promotes mental and physical health and increases its ability to withstand the effects of stressors like climate change. The promotion of green infrastructure, which is essential for resilience and health, is a key facet of the Healthy Cities Brisbane project [65]. For instance, Brisbane's massive Victoria Park transformation project is converting an old golf course into a sizable green area that will act as a natural flood mitigation system and offer leisure spaces that encourage socialisation and physical activity. In addition to aiding in the absorption of floodwaters, this natural solution cools the nearby metropolitan areas, lessening the risk of heat-related illness [26] and the urban heat island effect.

The program places a strong emphasis on active transport as well, understanding that encouraging healthier lifestyles requires bike lanes and walkability [27, 66]. By encouraging people to be more active and rely less on their cars, projects like the New Farm Riverwalk redevelopment have greatly improved access for bicyclists and pedestrians along Brisbane's riverfront, lowering emissions and improving air quality.

The Healthy Cities Brisbane initiative places a strong emphasis on community health hubs as a means of bringing healthcare closer to underserved areas and promoting resilience. Brisbane is attempting to lessen health disparities throughout its suburbs by advocating accessibility and placing healthcare services in community centers. In addition to providing residents with safe places and resources during heat waves or floods, these hubs also act as meeting places during extreme weather events. Additionally, the city is funding initiatives for community involvement and public awareness to motivate citizens to actively participate in creating a resilient and healthy city. Community gardens, mental health services, and workshops on climate adaptation are a few ways Brisbane is building a resilient, health-conscious populace [22] that can handle climate-related issues more effectively.

4 Discussion: Spatial Planning for Urban Resilience in Brisbane

4.1 Green Infrastructure and Flood Resilience

Brisbane's spatial planning initiatives now heavily rely on green infrastructure due to the growing demands of climate change, especially the frequency of floods. In addition to increasing urban resilience to environmental threats like flooding, green infrastructure—which includes parks, urban wetlands, green roofs, and other nature-based solutions—brings major health benefits to locals [22]. Brisbane can create areas that absorb excess water, reduce heat, enhance air quality, and provide chances for social and physical interaction by incorporating natural ecosystems into its urban areas. This all-encompassing strategy guarantees that the city's development is both resilient to climate change and health-promoting.

Brisbane frequently experiences flooding as a result of its subtropical climate and close proximity to significant river systems [28, 38]. Even though they are crucial, traditional gray infrastructure like stormwater drains and levees is frequently insufficient to handle the rising risk of extreme weather events. Conversely, green infrastructure offers a more adaptable and sustainable solution. To absorb rainwater, slow down runoff, and lessen the strain on stormwater systems, urban wetlands, parks, and permeable surfaces collaborate with nature [67]. This strategy lowers flood peaks, lessens the likelihood of flash flooding, and safeguards important urban infrastructure.

Urban wetlands are among the best types of green infrastructure for flood-prone areas. These ecosystems take up stormwater and gradually return it to the surroundings, acting as organic sponges [68–70]. By doing this, wetlands shield residences, commercial buildings, and vital infrastructure from excessive water flow and its rapidity. Furthermore, wetlands create healthier ecosystems, lower the risk of water-borne illnesses following floods, and filter pollutants and sediments to improve the quality of the water.

Urban parks and other green areas not only help to mitigate flooding but also have a substantial positive impact on public health. According to research, having access to green spaces improves mental health, lowers stress levels, and encourages physical activity. Parks provide areas for socialising, exercising, and playing, all of which are crucial for building a robust, healthy urban population [71]. Additionally, green spaces lessen the impact of the urban heat island, cooling the surrounding areas and lowering the risks associated with heat-related illnesses, especially in Brisbane during the hot summer months.

4.1.1 Case Study: Norman Creek Catchment Project

The Norman Creek Catchment Project is a prime example of how green infrastructure is being used in Brisbane to improve public health and flood resilience [21, 29]. Due to its natural topography and urban development, Norman Creek, which is only a few kilometers from Brisbane's central business district, has long been susceptible to flooding. Brisbane City Council launched the catchment project to regain the creek's natural functions and transform the surrounding area into a bustling community asset [72]. The project's integration of water-sensitive urban design (WSUD) principles is its central component. These guidelines center on controlling water flow, lowering runoff, and enhancing water quality through the use of the natural landscape [73]. Norman Creek restoration projects include rain gardens, bio-retention basins, and urban wetlands. By capturing stormwater and allowing it to gradually seep into the soil, these components lessen the chance of flash floods and ease the burden on the stormwater systems in the city. The wetlands and gardens contribute to the area's biodiversity by acting as animal habitats in addition to protecting against flooding.

The Norman Creek project has equally important health benefits. New public green spaces, bike and walking routes, and leisure areas that promote physical exercise and community involvement are all part of the restoration [74]. In an increasingly urbanized world, open green spaces provide chances for stress relief and relaxation, both of which are essential for mental health. Furthermore, the project has improved the resilience of adjacent neighbourhoods by lowering the risk of flooding, and providing a safer and more liveable environment for locals. Furthermore noteworthy as an illustration of community-driven resilience is the Norman Creek Catchment Project [74]. By actively participating in the project's design and execution, locals have made sure that it addresses their unique needs and promotes a feeling of environmental stewardship and ownership. Participation in the community builds social cohesion, which is important for urban resilience, especially in times of crisis.

4.2 Walkability and Active Living

A key component of urban planning that directly affects environmental sustainability and public health is walkability. Improving a neighbourhood's walkability in cities like Brisbane not only promotes more active lifestyles but also lowers carbon emissions by reducing the need for motor vehicles [75]. Building walkable environments promotes vibrant, healthy communities where people can easily access essential services, engage in physical activity, and enjoy a better quality of life—especially in urban areas that are becoming densely populated. Brisbane's New Farm Riverwalk redevelopment, which has turned the city's riverfront into a center for social interaction and active living [76], is one notable example of this.

Walkable urban neighbourhoods prioritise pedestrian mobility by providing convenient, safe, and accessible pathways for bicyclists and walkers. This strategy encourages people to walk or run errands, engage in recreational activities, or include physical activity in their everyday routines. The WHO states that preventing non-communicable diseases like obesity, diabetes, and heart disease requires regular physical activity [77]. Cities have the potential to significantly enhance public health outcomes by creating an appealing and simple walking option.

Additionally beneficial to mental health are walkable neighbourhoods. It has been demonstrated that having access to green areas and pedestrian-friendly streets lowers stress, anxiety, and depressive symptoms [78]. Even a short stroll through an urban park or along a street lined with trees can improve mood and cognitive performance. In today's fast-paced urban environments, where stress and mental health disorders are on the rise, the relationship between walkability and mental well-being is particularly significant.

Enhancing walkability in cities like Brisbane has substantial positive effects on environmental sustainability in addition to health benefits [79]. Walking and bicycling replace driving as a primary mode of transportation to help cut down on air pollution and greenhouse gas emissions. Brisbane's carbon emissions are mostly caused by transportation, with private automobiles playing a significant role [80]. The city can lessen its overall environmental impact, improve air quality, and lessen traffic congestion by establishing walkable areas [22]. Walkability promotes less energy-intensive, low-impact forms of transportation, which increases the city's resilience to climate change.

4.2.1 Case Study: New Farm Riverwalk Redevelopment as an Example of Fostering Active, Healthy Communities

Brisbane's New Farm Riverwalk redevelopment is a prime example of how enhancing walkability can promote vibrant, active neighbourhoods. The original Riverwalk, which ran alongside the Brisbane River from New Farm to the city center, was once an old and dilapidated building that washed away in the 2011 floods [81]. Understanding the value of this riverfront walkway, Brisbane City Council launched a significant reconstruction project to enhance the Riverwalk's accessibility, safety, and aesthetic appeal in addition to restoring it [82].

Since its reopening in 2014, the New Farm Riverwalk has gained popularity among joggers, cyclists, and walkers alike. The 850-m pedestrian and cycling path connects New Farm, a bustling inner-city suburb, to the central business district of the city while providing breathtaking views of the Brisbane River and city skyline [83]. It is a crucial piece of infrastructure that promotes active transportation and lowers the use of cars because of its well-planned layout and location.

The Riverwalk's layout provides roomy lanes for bicyclists and pedestrians, guaranteeing a secure and comfortable experience for all users. Its accessibility and appeal are further improved by the addition of seating, lighting, and shaded areas, which turn it into a place for socialising and exercise. In addition to supporting Brisbane's overarching objective of encouraging active living, the Riverwalk offers a beautiful, car-free route that encourages walking and cycling as viable options for daily commuting and leisure activities (Fig. 4).

The Riverwalk renovation's ability to lessen inner-city traffic congestion is one of its main advantages. The pathway that connects important neighbourhoods encourages people to walk or bike instead of driving, especially on short trips. This lessens the load on Brisbane's road system during peak hours in addition to lowering

Fig. 4 New farm river walk project. *Source* Maps are generated from publicly available maps on Google Earth Maps, generated on 21 October 2024

emissions. The Riverwalk provides a sustainable substitute for driving as Brisbane expands, assisting the city in meeting its carbon reduction goals.

The physical health of Brisbane's citizens has been positively impacted by the Riverwalk in terms of health benefits. The path is frequently used by cyclists and joggers, which promotes physical activity in the community. Furthermore, the communal nature of the Riverwalk, where locals congregate for walks, picnics, or workout sessions, fosters social cohesiveness and fortifies community bonds. Urban resilience depends on this feeling of community because it fosters the kind of support systems that are necessary during emergencies like floods or severe weather. Furthermore, future resilience was taken into consideration when designing the Riverwalk [83]. Built to withstand future flood events and elevated above the river, the structure reflects Brisbane's commitment to climate adaptation. The Riverwalk is a prime example of how urban infrastructure can be planned to reduce climate risk and promote health by combining resilience techniques with walkability.

4.3 Urban Farming and Community Health

Urban farming has the potential to improve public health, increase food security, and build community resilience, making it a cornerstone of sustainable urban living. Urban agriculture offers a means to re-establish residents' connection to local food production, lessen environmental impacts, and encourage healthier lifestyles in cities like Brisbane, where rapid urbanisation poses challenges to sustainability and public health issues. Urban farming is becoming recognised as a critical tactic to alleviate food insecurity, supply fresh produce, and promote social cohesion as Brisbane transitions to a more sustainable and health-conscious future. In the Kelvin Grove

Urban Village, for instance, community-based farming initiatives are opening up new avenues for locals [84, 85] to obtain sustainable diets and enhance their health.

Urban farming encompasses the practices of growing crops, raising livestock on a small scale, or gardening within urban areas. There are numerous ways to implement it, such as vertical gardens, rooftop farms, and community gardens. Urban farming offers an accessible, local solution to the problem of supplying fresh, healthful food as cities expand and encounter difficulties with their food supply chains. Urban farming is playing a bigger role in Brisbane, where climate-related risks like heatwaves and flooding affect food access and urban sprawl makes it harder to get food. This helps to ensure food security by lowering reliance on long-distance supply chains.

Improving food security is one of urban farming's most important contributions to Brisbane's urban environment. Urban farming lessens dependency on external food sources [86], which are subject to disruptions like extreme weather or problems with the global supply chain, by producing food locally. Urban farming guarantees a more stable and robust food system by enabling more people to grow their own food. This is especially important now that people are becoming more conscious of how lengthy supply chains affect society and the environment. Urban farms' close proximity also makes fresh food more readily available to locals, which minimises the need for long-distance transportation and carbon emissions.

Brisbane's urban farming is improving diets and ensuring food security at the same time. Fresh produce that is grown nearby motivates people to eat more plant-based foods, which are important in preventing lifestyle diseases like diabetes, obesity, and heart disease. Additionally, urban farming encourages people to be more conscious of their eating habits and the effects that their food choices have on the environment by strengthening the bond between them and their food. Community involvement is a common feature of Brisbane's urban farming initiatives, which improves social ties, lessens isolation, and fosters a sense of shared ownership over regional food systems [87]. Additionally, these initiatives offer educational opportunities, educating locals on environmental stewardship, nutrition, and sustainable farming methods. Urban farming contributes to healthier, more resilient urban communities through encouraging community involvement and education.

4.3.1 Case Study: Kelvin Grove Urban Village and Its Community-Based Farming Projects

One of Brisbane's best examples of the advantages of urban farming is the mixed-use Kelvin Grove Urban Village, which incorporates community gardens and urban farming into its architecture (Figs. 5 and 6). Kelvin Grove Urban Village, a creative, sustainable community that blends residential, commercial, and educational spaces with green areas and urban agriculture, is only two kilometers from Brisbane's central business district [88]. The village's emphasis on advancing health and wellbeing via neighbourhood-based farming initiatives is one of its distinguishing characteristics.

Urban farming at Kelvin Grove aims to foster sustainable lifestyles and a sense of community in addition to food production. The village has set up several community

Figs. 5 and 6 Kelvin Grove Urban Village in Brisbane in 2003 (above) and 2023 (below), showing two decades of development through community-driven farming projects. *Source* Maps are generated from publicly available maps on Google Earth Maps, generated on 21 October 2024

gardens where locals can cultivate their food, exchange gardening tips, and take part in group activities. Encouraging social interaction and intergenerational connections, these gardens are accessible to all, including families, students, and senior citizens. The cultivation of a diverse range of fruits, vegetables, and herbs in the community gardens at Kelvin Grove fosters a more sustainable diet by augmenting the availability of seasonal, fresh, and nutrient-dense produce. Because they can harvest their own food, locals are less likely to rely on produce from stores, which frequently has a larger environmental impact due to packaging and transportation. The practice of producing food locally also contributes to a decrease in food waste because locals are more inclined to eat what they grow and share any excess with their neighbours.

The health and happiness of the local community have been positively impacted by Kelvin Grove's urban farming initiatives. Because gardening requires physical labour and is good for both mental and cardiovascular health, the gardens give residents opportunities for physical activity. It has been demonstrated that gardening lowers stress, elevates mood, and fosters social interaction—all of which are essential for preserving mental health in urban settings. In addition, the gardens offer informative workshops on sustainable farming methods and nutrition, empowering locals to make better decisions regarding their diets and general health. The urban farming initiative in Kelvin Grove also makes a significant contribution to environmental sustainability [88]. Small-scale urban farming can lessen the environmental impact of food production, as the village's gardens show by encouraging the use of organic farming practices, composting, and water conservation measures. In addition to enhancing biodiversity and soil health, these methods lessen the carbon footprint left by conventional agriculture.

Brisbane's approach to sustainability, health, and food security is changing as a result of urban farming. The Kelvin Grove Urban Village serves as an example of how neighbourhood-based farming initiatives can enhance local access to wholesome, fresh food, encourage a healthier diet, and fortify social ties in urban areas [88]. Urban farming initiatives, such as the one at Kelvin Grove, provide a model for building resilient, health-focused communities that are better able to tackle the problems of food insecurity and climate change as Brisbane grows. Brisbane is helping its citizens live longer, healthier lives by incorporating urban farming into its city planning and development plans.

5 Concluding Remarks

This closing chapter has examined, from both global and local perspectives, how sustainable urban design can reinforce Brisbane's health-resilience nexus through a variety of case studies and insights. These examples, which range from community-based urban farming projects in Kelvin Grove Urban Village to walkability initiatives like the New Farm Riverwalk redevelopment, highlight the vital role that careful urban design plays in enhancing public health and community resilience. These projects offer a model for how urban spaces can be designed to meet the complex challenges of a rapidly changing world by promoting active living, enhancing food security, and reducing climate risks.

The case studies' key findings highlight a number of crucial lessons for Brisbane's future urban planning. Firstly, walkability is an important component of social interaction, mental health, and physical health in addition to being a transportation issue. Initiatives like the Riverwalk serve as examples of how putting pedestrian infrastructure first can lower emissions and promote active living at the same time. In a similar vein, Kelvin Grove Urban Village's urban farming demonstrates the importance of incorporating food production into urban areas, which improves food security, lessens environmental effects, and encourages community involvement. These programs offer real-world illustrations of how resilience and health can be incorporated into urban planning procedures.

Several policy recommendations emerge to further strengthen Brisbane's urban planning's connection between health and resilience. Initially, more attention ought to be paid to creating green infrastructure—parks, wetlands, and community gardens, for example—which not only enhances public health but also increases flood resilience and lessens the effects of urban heat islands. Furthermore, walkable neighborhoods should be given priority in Brisbane's urban planning policies, and funds for accessible, safe pathways for bicyclists and pedestrians should be allocated, particularly in the outer suburbs where such infrastructure is frequently lacking. Additionally, integrating health impact assessments into urban development procedures will guarantee that public health continues to be a top priority in planning for the future.

Assessing the long-term health effects of sustainable urban design interventions should be the main goal of future research in the health-resilience field. Studies that monitor the long-term effects of walkability and urban farming initiatives on mental and physical health may fall under this category. How various communities—particularly marginalised ones—experience and benefit from these interventions in urban design is a crucial topic of research. Ensuring that health-resilience initiatives are equitable and accessible to all residents is crucial for achieving broader sustainability goals.

References

1. Cheshmehzangi A (2016) Multi-spatial environmental performance evaluation towards integrated urban design: a procedural approach with computational simulations. J Clean Prod 139:1085–1093
2. Cheshmehzangi A, Zhu Y, Li B (2010) Integrated design approach-urban design for sustainability. In: At the 5th international conference on responsive manufacturing—green manufacturing (ICRM 2010), pp 241–247. Available from: https://doi.org/10.1049/cp.2010.0441
3. Barton H (2009) Land use planning and health and well-being. Land Use Policy 26:S115–S123
4. Barton H (2016) City of well-being: a radical guide to planning. Routledge, New York
5. Brown K, Westaway E (2011) Agency, capacity, and resilience to environmental change: lessons from human development, well-being, and disasters. Annu Rev Environ Resour 36(1):321–342
6. Mehmood A (2016) Of resilient places: planning for urban resilience. Eur Plan Stud 24(2):407–419
7. Taylor L, Hochuli DF (2015) Creating better cities: how biodiversity and ecosystem functioning enhance urban residents' wellbeing. Urban Ecosyst 18:747–762
8. Zautra AJ, Arewasikporn A, Davis MC (2010) Resilience: promoting well-being through recovery, sustainability, and growth. Res Hum Dev 7(3):221–238
9. Celentano G, Obonyo E, Habert G (2023) Achieving social justice and environmental justice in safe affordable housing through a materials-centered, multi-level, transdisciplinary approach. In: The Routledge handbook of embodied carbon in the built environment. Routledge, New York, pp 275–295
10. Luginaah I, Kerr RB (eds) (2015) Geographies of health and development. Ashgate Publishing Ltd., London
11. Castleden M, McKee M, Murray V, Leonardi G (2011) Resilience thinking in health protection. J Public Health 33(3):369–377
12. Datola G (2023) Implementing urban resilience in urban planning: a comprehensive framework for urban resilience evaluation. Sustain Cities Soc 98:104821

13. WHO (World Health Organization) (2001) The world health report 2001: mental health: new understanding, new hope. Available from: https://iris.who.int/handle/10665/42390
14. Beck MB, Villarroel Walker R (2013) Nexus security: governance, innovation and the resilient city. Front Environ Sci Eng 7:640–657
15. Dodman D, Diep L, Colenbrander S (2017) Making the case for the nexus between resilience and resource efficiency at the city scale. Int J Urban Sustain Develop 9(2):97–106
16. Parihar J, Birman S (2024) Heat resilience in urban environments: strategies for sustainable city climate management. In: The Climate-health-sustainability nexus: understanding the interconnected impact on populations and the environment. Springer, Cham, pp 305–324
17. Lynch A, Nicholls N, Alexander L, Griggs D (2008) Defining the impacts of climate change on extreme events. Garnaut Clim Change Rev 2008:1–19
18. Tira M, Pezzagno M, Richiedei A (eds) (2020) Pedestrians, urban spaces and health: proceedings of the XXIV international conference on living and walking in cities (LWC, September 12–13, 2019, Brescia, Italy). CRC Press, Boca Raton
19. Brockie L, Miller E (2017) Understanding older adults' resilience during the Brisbane floods: social capital, life experience, and optimism. Disaster Med Public Health Prep 11(1):72–79
20. Irajifar L, Sipe N, Alizadeh T (2016) The impact of urban form on disaster resiliency: a case study of Brisbane and Ipswich, Australia. Int J Disaster Resilience Built Environ 7(3):259–275
21. Zaman KUAB, Chowdhooree I (2022) Water sensitive urban design for enhancing flood resilience: a case study in Brisbane city, Australia. In: Handbook of waterfront cities and urbanism. Routledge, New York, pp 145–162
22. Cheshmehzangi A (2024) Why Brisbane needs more green space, not less, an opinion piece in Brisbane Times. Available from: https://www.brisbanetimes.com.au/national/queensland/why-brisbane-needs-more-green-space-not-less-20240904-p5k7x9.html
23. Rahman R, Zafarullah H (2020) Impact of climate change on human health: adaptation challenges in Queensland, Australia. Clim Res 80(1):59–72
24. Arthurson K, Baum S (2015) Making space for social inclusion in conceptualising climate change vulnerability. Local Environ 20(1):1–17
25. Astell-Burt T, Feng X, Mavoa S, Badland HM, Giles-Corti B (2014) Do low-income neighbourhoods have the least green space? A cross-sectional study of Australia's most populous cities. BMC Public Health 14:1–11
26. Cheshmehzangi A, Butters C (2014) The effects of improved built form on urban cooling: review of passive strategies and urban design case studies. SEE 2014, Bangkok
27. Bajracharya B, Khan S (2020) Urban governance in Australia: a case study of Brisbane City. New Urban Agenda Asia Pac Gov Sustain Inclusive Cities 2020:225–250
28. Brage T, Leardini P (2018) Planning opportunities for climate resilience in Brisbane, Australia: learning from The City's History of Floods. WIT Trans Built Environ 184:91–103
29. Tan J (2021) Blue-sky thinking: flood risk management and blue–green infrastructure in Brisbane and Singapore. Master degree thesis, University of Otago, Dunedin, New Zealand. Available from: https://ourarchive.otago.ac.nz/view/pdfCoverPage?instCode=64OTAGO_INST&filePid=13397354050001891&download=true
30. Wickes R, Zahnow R, Taylor M, Piquero AR (2015) Neighborhood structure, social capital, and community resilience: longitudinal evidence from the 2011 Brisbane flood disaster. Soc Sci Q 96(2):330–353
31. Cheshmehzangi A (2022) Towards sustainable urban green infrastructures. In: Green infrastructure in Chinese cities. Springer, Singapore, pp 495–505
32. Brisbane City Council (BCC) (2019a) The original CityShape 2026. Available from: https://www.brisbane.qld.gov.au/planning-and-building/planning-guidelines-and-tools/neighbourhood-planning-and-urban-renewal/neighbourhood-planning-and-urban-renewal-process/brisbane-cityshape-2026
33. Shill J, Busst C, Horton K, Corben K, Demaio S (2021) Our path to health for all: Australia in 2030. Med J Aust 214(8 Supplement):5–6
34. Lamond J, Stanton-Geddes Z, Bloch R, Proverbs D (2013, May) Cities and flooding: lessons in resilience from case studies of integrated urban flood risk management. In: CIB 2013 world congress, special conference session on making cities more resilient, Brisbane, Australia

35. Loughnan M, Tapper N, Phan T (2014) Identifying vulnerable populations in subtropical Brisbane, Australia: a guide for heatwave preparedness and health promotion. Int Schol Res Not 2014(1):821759
36. Adame MF, Fry B, Bunn SE (2016) Water isotope characteristics of a flood: Brisbane River, Australia. Hydrol Process 30(13):2033–2041
37. Deo RC, Byun HR, Adamowski JF, Kim DW (2015) A real-time flood monitoring index based on daily effective precipitation and its application to Brisbane and Lockyer Valley flood events. Water Resour Manag 29:4075–4093
38. Smith I, McAlpine C (2014) Estimating future changes in flood risk: case study of the Brisbane River, Australia. Clim Risk Manag 6:6–17
39. Erickson TB, Brooks J, Nilles EJ, Pham PN, Vinck P (2019) Environmental health effects attributed to toxic and infectious agents following hurricanes, cyclones, flash floods and major hydrometeorological events. J Toxicol Environ Health Part B 22(5–6):157–171
40. Kakalou E, Tsiamis C (2021) Infectious diseases outbreaks following natural disasters: risk assessment, prevention, and control. In: Medicine E (ed) Trauma and disaster management: from prehospital to hospital care and beyond. Springer, Cham, pp 525–535
41. Kuutti-Hintikainen A (2019) Health risks as a result of floods caused by climate change: the Finnish authorities preparedness. Diakonia University of Applied Sciences Master's Degree in Global Health Care Master of Global Health Care. Available from: https://www.theseus.fi/handle/10024/265226
42. Pernice R, Chen B (eds) (2024) Australia and China perspectives on urban regeneration and rural revitalization. Taylor & Francis, Milton Park
43. Baum S (2008) Suburban scars: Australian cities and socio-economic deprivation (No. 15). Griffith University, Urban Research Program. Available from: https://citeseerx.ist.psu.edu/document?repid=rep1&type=pdf&doi=3ae1cfc9d7e6bb19378817be4f692b4b58ff27c0
44. Ross CE, Mirowsky J (2001) Neighborhood disadvantage, disorder, and health. J Health Soc Behav 2001:258–276
45. Department of Health (2019) 2018–2019 annual report. Available from: https://www.health.qld.gov.au/__data/assets/pdf_file/0019/882010/190927-DoH-Annual-Report-2018-19.pdf
46. Queensland Health (2018) 2018 Queensland health report. Available from: https://www.health.qld.gov.au/__data/assets/pdf_file/0032/732794/cho-report-2018-full.pdf
47. AIHW (Australian Institute of Health and Welfare) (2021) Australia's welfare 2021: in brief. Available from: https://www.aihw.gov.au/reports/australias-welfare/australias-welfare-2021-in-brief/contents/summary
48. Queensland Government (2023) Mental health. Available from: https://www.choreport.health.qld.gov.au/our-health/mental-health
49. Callaghan J (2023) The Eastern Australian Floods of February 2022 and its relationship with climate change. J Energy Power Technol 5(2):1–42
50. Levin N, Phinn S (2022) Assessing the 2022 flood impacts in Queensland combining daytime and nighttime optical and imaging radar data. Remote Sens 14(19):5009
51. Vardoulakis S, Matthews V, Bailie RS, Hu W, Salvador-Carulla L, Barratt AL, Chu C (2022) Building resilience to Australian flood disasters in the face of climate change. Med J Aust 217(7):342
52. WHO (World Health Organization) (2015) Healthy cities. Available from: https://www.who.int/teams/health-promotion/enhanced-wellbeing/ninth-global-conference/healthy-cities
53. WHO (World Health Organization) (1986) Creating healthy cities. Available from: https://www.who.int/activities/creating-healthy-cities
54. WHO (World Health Organization) (2020) Healthy cities effective approach to a rapidly changing world. World Health Organization. Available from: https://www.who.int/publications/i/item/9789240004825
55. WHO (World Health Organization) (2023) Guidance on engagement of communities and civil society to end tuberculosis. World Health Organization. Available from: https://www.who.int/publications/i/item/9789240080294

56. Cheshmehzangi A, Butters C, Cheshmehzangi A, Dawodu A, Jin R, Mangi E (2022). Urban design in the age of climate change: paradigms and directions. In: Design for vulnerable communities. Springer, Cham, pp 133–154
57. Cheshmehzangi A, You N, Siri J, Birch E (2024) Harnessing urban innovation to unlock the sustainable development goals. Springer, Singapore
58. Zou T, Dawodu A, Mangi E, Cheshmehzangi A (2023) Exploring current trends, gaps & challenges in sustainable food systems studies: the need of developing urban food systems frameworks for sustainable cities. Sustainability 15(13):10248
59. Boychev M (2021) Reshaping cities through sustainable urban planning strategies and creating healthy, equitable and sustainable communities addressing the united nations sustainable development goals. Available from: https://yorkspace.library.yorku.ca/server/api/core/bitstreams/f114b832-feda-4305-a380-647c82e476fd/content
60. Lowe M, Arundel J, Hooper P, Rozek J, Higgs C, Roberts R, Giles-Corti B (2020) Liveability aspirations and realities: implementation of urban policies designed to create healthy cities in Australia. Soc Sci Med 245:112713
61. Cheshmehzangi A, Griffiths CJ (2014) Development of green infrastructure for the city: a holistic vision towards sustainable urbanism. J Archit Environ 2(2):13–20
62. Council of Australian Governments (2011) National strategy for disaster resilience. Available from: https://www.homeaffairs.gov.au/emergency/files/national-strategy-disaster-resilience.pdf
63. Biswas RR, Rahman A (2023) Adaptation to climate change: a study on regional climate change adaptation policy and practice framework. J Environ Manag 336:117666
64. Department of Environment and Heritage Protection (2017) Queensland climate adaptation strategy 2017–2030, pathways to a climate resilient Queensland. Available from: https://www.qld.gov.au/__data/assets/pdf_file/0017/67301/qld-climate-adaptation-strategy.pdf
65. Brisbane City Council (BCC) (2024) Brisbane. Clean, green, sustainable. Available from: https://www.brisbane.qld.gov.au/sites/default/files/documents/2024-02/20240202-Brisbane-Clean-Green-Sustainable.PDF
66. Fishman E, Nahseri M, Davies L, Tran J, Katsu T (2024) Prioritising active transport (no. AP-R711-24). Available from: https://austroads.com.au/__data/assets/pdf_file/0031/653368/AP-R711-24_Prioritising_Active_Transport.pdf
67. Cheshmehzangi A, Sedrez M, Flynn A (2024) Aligning with stormwater management but from the urban design perspective: towards sustainability-driven and water-based planning and design. In: Rethinking stormwater management through sustainable urban design. Springer, Singapore, pp 319–327
68. Li L, Chan F, Cheshmehzangi A (2023) Nature-based solutions and sponge city for urban water management. In: Adapting the built environment for climate change. Woodhead Publishing, pp 371–402
69. Mohamed NMM (2024) Constructed wetland systems a proactive model for ecosystems & flood protection services (Case study: Egyptian Mediterranean urban communities). Doctoral dissertation, Cairo University, Egypt
70. Sun J, Cheshmehzangi A, Wang S (2022) Comprehensive evaluation of green infrastructure restorative practices for high-quality transitional "sponge node" renewal programs in China. In: Green infrastructure in Chinese cities. Springer, Singapore, pp 265–298
71. Ziafati Bafarasat A, Cheshmehzangi A, Ankowska A (2023) A set of 99 healthy city indicators for application in urban planning and design. Sustain Dev 31(3):1978–1989
72. Cook MH (2023) A river with a city problem: a history of Brisbane floods. University of Queensland Press (UQP), Brisbane
73. Leonard R, Walton A, Koth B, Green M, Spinks A, Myers B, Malkin S, Mankad A, Chacko P, Sharma A, Pezzaniti D (2014) Community acceptance of water sensitive urban design: six case studies. Goyder Institute for Water Research, Adelaide, Australia. Available from: https://goyderinstitute.org/wp-content/uploads/2023/03/goyder_trs_14-3_community_acceptance_wsud.pdf

74. Brisbane City Council (BCC) (2024) Noman creek 2012–2031 project. Available from: https://www.brisbane.qld.gov.au/clean-and-green/natural-environment-and-water/creeks-and-waterways/protect-our-waterways/norman-creek-2012-2031-project
75. Hellberg R, Guaralda M, Rinchumphu D (2021) Urban walkability profiles in Brisbane. Int Rev Spat Plan Sustain Develop 9(3):1–15
76. Queensland Government (2020) Brisbane Ferry terminals and Riverwalk rebuilt. Available from: https://www.qra.qld.gov.au/news-case-studies/case-studies/brisbane-ferry-terminals-and-riverwalk-rebuilt
77. WHO (World Health Organization) (2024) Physical activity. Available from: https://www.who.int/news-room/fact-sheets/detail/physical-activity
78. Senthil M (2024) Exploring the impact of psychopathology spaces in urban environments: towards improved living and urban stress management. Int J Innov Res Multidiscip Field 10(4):94–101. https://doi.org/10.2015/IJIRMF/202404015
79. Gunn LD, Mavoa S, Boulangé C, Hooper P, Kavanagh A, Giles-Corti B (2017) Designing healthy communities: creating evidence on metrics for built environment features associated with walkable neighbourhood activity centres. Int J Behav Nutr Phys Act 14:1–12
80. McGreevy M, Harris P, Delaney-Crowe T, Fisher M, Sainsbury P, Riley E, Baum F (2020) How well do Australian government urban planning policies respond to the social determinants of health and health equity? Land Use Policy 99:105053
81. Cook MH (2018) A river with a city problem, not a city with a river problem': Brisbane and its flood-prone river. Environ Hist 24(4):469–496
82. Brisbane City Council (BCC) (2019b) River's edge strategy. Available from: https://www.brisbane.qld.gov.au/planning-and-building/planning-guidelines-and-tools/neighbourhood-planning-and-urban-renewal/neighbourhood-plans-and-other-local-planning-projects/river-access-network/rivers-edge-strategy
83. Bricycle (n.d.) Riverwalk floating walkway. Available from: https://briscycle.com/brisbane-city/riverwalk/
84. Crane P, Wyeth S, Brough M, Spencer A (2006) Children in inner city suburbia: the case of new farm, Brisbane. In: Second national conference on creating child-friendly cities. Australian Research Alliance for Children and Youth, pp 1–37
85. Crombie RC, Delaforce WH (2008) Kelvin grove urban village pathways audit report. Available from: https://eprints.qut.edu.au/13591/
86. Martin-Moreau M, Ménascé D (2019) Urban agriculture: another way to feed cities. Field actions science reports, 20. Available from: https://journals.openedition.org/factsreports/pdf/5536/1000
87. Summerhayes L, Baker D, Vella K (2024) Food diversity and accessibility enabled urban environments for sustainable food consumption: a case study of Brisbane, Australia. Human Soc Sci Commun 11(1):1–14
88. Aurecon (n.d.) Kelvin grove urban development: designing Australia's first master planned urban development, Available from: https://www.aurecongroup.com/projects/government/kelvin-grove-urban-development